Progress in Theoretical Computer Science

Editor
Ronald V. Book, University of California

Editorial Board
Erwin Engeler, ETH Zentrum, Zurich, Switzerland
Gérard Huet, INRIA, Le Chesnay, France
Jean-Pierre Jouannaud, Université de Paris-Sud, Orsay, France
Robin Milner, University of Edinburgh, Edinburgh, Scotland
Maurice Nivat, Université de Paris VII, Paris, France
Martin Wirsing, Universität Passau, Passau, Germany

Heinrich Hussmann

Nondeterminism in Algebraic Specifications and Algebraic Programs

Birkhäuser

Boston · Basel · Berlin

Heinrich Hussmann
Technische Universität München
Institut für Informatik
D-8000 München 2
Germany

Library of Congress Cataloging-in-Publication Data

Hussmann, Heinrich, 1959-
    Nondeterminism in algebraic specification and algebraic programs /
Heinrich Hussmann.
        p.   cm. -- (Progress in theoretical computer science)
    Includes bibliographical references,
    ISBN-13: 978-1-4684-6836-6   e-ISBN-13: 978-1-4684-6834-2
    DOI: 10.1007/978-1-4684-6834-2
    1. Computer science--Mathematics. I. Title. II. Series.
  QA76.9.M35H87   1993   93-9340
    005.13'1--dc20   CIP

Printed on acid-free paper.

© Birkhäuser Boston 1993.
Softcover reprint of the hardcover 1st edition 1993

ISBN-13: 978-1-4684-6836-6

Typset copy prepared by the Author.

9 8 7 6 5 4 3 2 1

Table of Contents

Preface

Algebraic specification, nondeterminism and term rewriting are three active research areas aiming at concepts for the abstract description of software systems: Algebraic specifications are well-suited for describing data structures and sequential software systems in an abstract way. Term rewriting methods are used in many prototyping systems and form the basis for executing specifications. Nondeterminism plays a major role in formal language theory; in programming it serves for delaying design decisions in program development and occurs in a "natural" way in formalisations of distributed processes.

Heinrich Hussmann presents an elegant extension of equational specification and term rewriting to include nondeterminism. Based on a clean modeltheoretic semantics he considers term rewriting systems without confluence restrictions as a specification language and shows that fundamental properties such as the existence of initial models or the soundness and completeness of narrowing, the basic mechanism for executing equational specifications, can be extended to nondeterministic computations.

The work of Heinrich Hussmann is an excellent contribution to Algebraic Programming; it gives a framework that admits a direct approach to program verification, is suitable for describing concurrent and distributed processes, and it can be executed as fast as Prolog.

Munich, January 1993   Martin Wirsing

Preface by the Author

This monograph is based on a Ph. D. thesis with the title „Nichtdeterministische Algebraische Spezifikationen" (in German), which was accepted by the University of Passau in the winter term 1988/1989. The text has been thoroughly revised and substantially extended in the meantime. The original thesis was primarily aimed at the "core" theme of generalizing algebraic specifications to nondeterminism. The monograph version now contains a comparison with the established field of logic programming, a new chapter on narrowing for nondeterministic specifications and a more detailed treatment of graph rewriting techniques. I hope that the revision will extend the readership of this book from specialists in algebraic specification to everybody who is interested in the relationship between logic programming, term rewriting and formal specification.

I would like to thank Prof. Broy for finding the topic of the thesis and for his manyfold support, in particular for many fruitful discussions and for reading a preliminary version. Many thanks also to Prof. Wirsing, for reading a draft, for valuable remarks and for proposing the publication within this series of books. The anonymous referee has given very helpful suggestions for the revision of the manuscript, many thanks also to him or her.

Many collegues have contributed to this work by discussions and remarks; for their particular interest I would like to thank Thomas Belzner, Alfons Geser, Jürgen Knopp, Bernhard Möller, Peter Mosses, Tobias Nipkow, Gert Smolka and Michal Walicki.

Gabi Kohlmaier, now Gabi Hussmann, is entitled to special warm thanks for moral support.

Finally, I would like to thank the staff of Birkhäuser for their excellent support in the technical production of the book.

Munich, December 22, 1992   Heinrich Hussmann

Chapter 0

Introduction

This monograph presents a generalization of the theory of equational algebraic specifications, where the equational axioms are replaced by directed rewrite rules. A model-theoretic semantics for such specifications is given, which provides a rather general framework for studying

- the integration of nondeterminism into algebraic specifications, and
- model-oriented semantics for general (non-confluent) term rewriting.

The study of this central topic leads to interesting side results in the fields of

- relationships between algebraic and logic programming, and
- relationships between term rewriting and graph rewriting.

The starting point for this work is the observation that the available formal specification languages for software are very much influenced by the concepts of traditional mathematical logic. In particular, the notion of equality (which is a *symmetric* operation) plays a central role in algebraic specifications. This emphasis on symmetry does not correspond well to the fact that software belongs to a computational paradigm, which is always *directed*. Every execution of an algorithm consists in a directed evaluation of its formal descripton (therefore leading to such problems as the question of termination). This kind of directed evaluation transforms syntactical objects into *semantically equal* ones. Classical (deterministic) evaluation gives a close connection between a non-symmetric relation between objects (the operational evaluation) and a symmetric one (the semantical denotation). The theory of term rewriting is an ideal framework for studying such connections.

The central step during an evaluation using a term rewrite system needs two decisions: which rewrite rule to apply and at which position (redex) of the actual target term. This obviously is a situation of *nondeterminism*. (Nondeterminism means here that the next step within the computation is not uniquely determined.) The whole well-developed theory of canonical term rewrite systems is concerned with conditions which ensure that this implicit nondeterminism does not affect the computation of the actual result. In other words it uses the directed relation operationally, but simultaneously keeps the symmetrical relation between objects on the semantical level.

The basic idea of the approach presented here is to make positive use of the powerful framework of term rewriting, as a tool to specify nondeterministic computations. It is a rather special (and, admittedly, important) subcase where a nondeterministic computation is determined to give one fixed result. However, there are many situations in computing where nondeterminism is explicitly present or even needed. The most typical occurrences of nondeterminism are:

- Distributed systems, using concurrency and communication. Here, the actual result of a computation depends on various parameters (including messages from other components), which are not completely predictable. So it is necessary to classify a *set of possible results* of a particular computation.
- Stepwise program development. In an abstract description of a program, it is often useful to keep a whole range of implementations open using a nondeterministic choice like "Choose an arbitrary element of the set M".

These are also the reasons why much of the research even in the early days of computer science was invested into the investigation of nondeterminism.

So the general goal for this text is to employ the formalism for nondeterministic computations, which is given by (non-canonical) term rewriting, as a specification language. In difference to classical term rewriting, the interpretation of a rewriting step is no longer the (symmetric) semantic equivalence but a directed notion. It is quite obvious that the appropriate directed notion on the semantical level is *set inclusion*, since every state of a nondeterministic computation in fact describes a *set* of possible results, and every step of computation can make a *choice*, which further restricts the set of possible results. The concept of a *multi-algebra*, that is an algebra where the operations are interpreted by *set-valued* functions, gives the appropriate semantical background for such an interpretation.

These considerations give a clear technical working plan for this book:

(1) Give a detailed technical definition for the "set-valued" interpretation sketched above.

(2) Investigate soundness and completeness of term rewriting with respect to the semantics given in step (1).

(3) Demonstrate that equational algebraic specifications form a subcase of the new approach.

(4) Investigate the integration of the new approach with advanced concepts of algebraic specification and term rewriting, like conditional rules, partial algebraic specifications, or theory-unification procedures.

(5) Compare the new approach with other frameworks, like logic programming.

From the traditional theory of algebraic specifications, another point comes onto the working plan, which is somehow related to (2):

(2a) Investigate the structure of model classes, in particular the existence of initial models.

The text provides results for all the steps of the working plan. However, at various stages, the syntax of specifications and also the rewriting calculus have to be enriched and adapted in order to get sensible results.

Before going into the technical details, the next section gives a sketchy overview of the main stream of argumentation, and puts together the main results presented in this monograph.

0.1 Preview

The notion of equational algebraic specifications is generalized to nondeterministic specifications. As usual, a specification consists of a signature (defining sort and function symbols) and a set of axioms. Syntactically, the main difference to classical specifications is that the axioms now contain the directed symbol "→" instead of the symmetric symbol "=".

Semantically, a class of *multi-algebras* is associated with a given specification. Within a multi-algebra A, a function symbol

$$f: s1 \to s2$$

is interpreted by a *set-valued* function

$$f^A: s1^A \to \wp^+(s2^A),$$

which delivers a finite and non-empty subset of $s2^A$ as its result. It is important to note that for the interpretation of a term, the set-valued functions are put together in an *additive* way. For instance (let $I^A[t]$ denote the interpretation of term t in multi-algebra A):

$$I^A[\, f(g)\,] = \{\, e \in f^A(e') \mid e' \in g^A \,\}.$$

This gives an interpretation of terms by sets of elements from the carrier sets. The interpretation is easily extended to terms with variables, using an environment β which provides actual values for the variables.

The validity of a (directed) axiom is now given by set inclusion: An axiom

$$t1 \to t2$$

is valid in a multialgebra A, iff for all environments β we have

$$I_\beta^A\, [t1] \supseteq I_\beta^A\, [t2].$$

This concludes (apart from technical details) the step (1) of the working plan.

For step (2), soundness and completeness of standard term rewriting have to be investigated with respect to the new semantics. This leads to a negative result: standard term rewriting is *unsound* in this sense. The reason for this is closely related to procedure call conventions in programming languages. The term rewriting approach uses a technique similar to "call-by-name", whereas the multi-algebras have a clear "call-by-value" semantics. (In the detailed exposition, the more precise terms "run-time choice" and "call-time choice" will be used.) This makes a difference for rewrite rules where a variable occurs several times in the right hand side, like in:

$$\text{double}(x) \to \text{add}(x,x)$$

If the variable x is substituted by a nondeterministic term, then term rewriting generates *two independent copies* of the term in the right hand side which can be evaluated separately to different values. However, in the interpretation of the axioms for a model, the interpretation of the variable x is a single value, which is *the same for all occurrences* of x. So the term rewriting process deduces consequences from the axioms which are not semantically valid in all models. This is a first (negative) result:

Standard term rewriting and multi-algebras rely on different semantic concepts.

One way to overcome this difficulty is to introduce a second kind of axioms which gives a syntactic way to state that a term must be interpreted deterministically. Such axioms (determinacy axioms) are written like

DET(t)

which means: "The interpretation of t in every model must be a one-element set."

The term rewriting calculus now can be adapted to this concept. It is now only allowed to substitute a term for a variable of an axiom, if the substituted term has been proven to be deterministic. This gives the calculus a much more "call-by-value" flavour (and differs from standard term rewriting). We call this new calculus "DET-rewriting" here, for short. It turns out that DET-rewriting is sound, but unfortunately it can be shown now to be *incomplete*. This is a second (negative) result:

The introduction of determinacy rules into specifications and calculus achieves soundness, but does not suffice to ensure completeness of the calculus.

The reason for the problem can be understood best when looking at an attempt to constuct a term model for a specification (which is the usual technique to prove completeness). A specification may contain the following axioms:

$$f(g) \to a, \qquad g \to b, \qquad f(a) \to b, \qquad f(b) \to b,$$
$$DET(a), \qquad DET(b).$$

A term model should basically interpret every term by the set of deterministic terms it can be reduced to within the calculus. If the interpretation of the term $f(g)$ in such a term model is built up piecewise from the operations, the set { b } is the natural result (since g can be reduced to b only). However, the axioms require the interpretation of $f(g)$ to contain a, too.

The problem with axioms like ⟨$f(g) \to a$⟩ above is that they do not admit an additive construction of a term model. So they are excluded by a syntactical condition for axioms, which is called *DET-additivity*. DET-additivity is a rather complex condition, which fortunately can be ensured by simple syntactical criteria. A simple and useful criterion is that in the left hand side of an axiom only the topmost symbol is allowed to be a nondeterministic operation.

Under the precondition of DET-additivity, soundness and completeness results hold. In general, only *weak ground* completeness can be shown, which means that every logically valid inclusion ‹t1 → t2› is deducible in the case, where t1 and t2 do not contain variables and t2 is a deterministic term. This is a (positive) result:

> *Under the precondition of DET-additivity, DET-rewriting is sound and weakly ground complete.*

The book contains more detailed investigations how to achieve more general completeness results, which are not reflected here. An interesting side effect of the completeness proof is that the constructed term model turns out to be an initial one.

> *Under the precondition of DET-additivity, initial models always exist.*

Stepping to item (3) of the working plan from above, it can be easily shown that classical equational specifications are a subcase of the new approach (by simply declaring all terms as deterministic).

> *Equational specifications are a special case of nondeterministic specifications.*

Step (4) of the working plan contains several, mainly unrelated pieces, most of which do not uncover unexpected effects. One topic which causes rather diffcult technicalities is the integration of partial functions. The combination of nondeterminism with partiality is slightly problematic, but an approach can be found which generalizes the results in a satisfactory manner to partiality.

> *The main results can be carried over to the case of partial functions.*

In order to investigate some questions related to the steps (4) and (5), an interesting and important subclass of nondeterministic algebraic specifications is identified, which are called *constructor-based*. This is inspired by a very popular style of algebraic specfications. Basically, a subset of the function symbols is designated as the so-called constructors, and the left hand sides of the rules are restricted to terms of the shape

$$f(c_1, ..., c_n)$$

where f is a non-constructor function symbol, and the c_i consist only of constructor symbols and variables. In contrast to classical algebraic specifications, we do not require here any conditions besides this syntactic one. In particular, the "principle of definition" is *not* assumed (which for instance would require a complete case analysis over all constructors for any non-constructor symbol to be given). If constructors are assumed to be deterministic (and always defined), then such specifications are automatically DET-additive.

> *Constructor-based specifications are characterized only syntactically. All results hold for constructor-based specifications without any additional precondition.*

For this special class of specifications, also the extension to a unification algorithm can be given successfully. There is a complete narrowing procedure for such nondeterministic specifications. This completeness holds independently of confluence or termination of the rule system, giving an interesting generalization of results in the literature.

> *For constructor-based nondeterministic specifications, narrowing is complete without any confluence or termination conditions.*

This result builds the bridge to step (5) of the working plan, which examines the connections to logic programming. The special case of constructor-based nondeterministic specifications can be shown to be one-to-one related to logic programming for definite clauses. This also generalizes results known from the literature, which needed the restriction to canonical rewrite systems.

> *For constructor-based nondeterministic specifications, there is a one-to-one correspondence between narrowing and logic programming.*

Finally, a new point (6) appears on the working plan, which has not been mentioned above. Since the DET-rewriting calculus differs from standard term rewriting, it is questionable, whether existing implementations of term rewriting can be used for the new approach. Fortunately, for the subcase of constructor-based specifications a positive result can be found. If an implementation of term rewriting is used which represents terms by directed acyclic graphs with "variable sharing", the implementation is sound and complete with respect to the DET-rewriting calculus. A particularly interesting observation is that the "sharing" of subterms used in such implementations takes care of the soundness with respect to multiple occurrences of variables. So an arbitrary redex selection strategy again is admissible, as in standard term rewriting.

> *For constructor-based nondeterministic specifications, any implementation of term rewriting using shared term structures is sound and complete with respect to the multi-algebra semantics.*

Thus, for the special case of constructor-based specifications, the original aims are reached completely, despite of the discouraging negative results during the first steps of the study.

The book is structured as follows: The second section of this introductory chapter (chapter zero) gives an overview of the historical background of this work, a third section introduces some basic notions. The subsequent first chapter already contains the elementary framework for nondeterministic algebraic specifications (the notion of a model and the calculus of term rewriting). Within this chapter, emphasis is put on a detailed motivation for the design decisions. The definition of the calculus leads to complications which can be resolved in a second version of the theory, presented in the second chapter. In the third chapter, the particular question of a lattice structure of the model classes is dealt with, a topic which may be skipped by the reader not interested in semantic investigations. At this point the theory has gained some kind of completeness, so the fourth chapter gives a detailed view of the relationships between the new approach presented here and other approaches such as equational logic, term rewriting, and logic programming. The fifth chapter concerns itself with more practical consequences: In its first part it deals with implementation issues from a rather abstract point of view; in its second part the application to a number of simple examples is demonstrated, taken from various areas of computer science. The sixth chapter again treats theoretical questions, and that is to integrate the new approach with a treatment of partial functions, as proposed in [Broy, Wirsing 82]. The concluding (seventh) chapter shows the application of nondeterministic algebraic specifications to a non-trivial example: the language of communicating sequential processes.

0.2 Historical Background

The technique of *algebraic specifications*, established in the years 1975–80 ([ADJ78], [Guttag 75]), is an attempt to use results of Universal Algebra ([Birkhoff35]) for the mathematical description of data structures ("abstract data types"). The basic idea of this approach is to describe a data domain together with its characteristic operations. An algebraic specification has a precise mathematical semantics, given by its *models* (which are so-called *heterogeneous algebras*, consisting of data sets and operations on them). The specification restricts the class of models by a number of *axioms* in a logic language. So there is a corresponding *calculus* which admits to *derive* further properties of a specification. Of particular interest is the *evaluation* of expressions over the specification which is an abstract form of operational semantics. These basic ideas have been refined in various ways, for instance by concepts for modularisation and the treatment of partial functions ([Wirsinget.al. 83]). Altogether, a specification language arose which combined the expressive power of a programming language with a formal treatment of data types.

Even earlier, about 1970–75, *nondeterminism* has been recognized as important for the abstract description of programs ([Floyd 67], [Manna70], [Dijkstra76]). Up to now it is an open question whether nondeterminism is useful for practical programming. But as it was argued above, a demand of *abstractness* within descriptions often leads to nondeterminism. Although abstractness was the aim of algebraic spefication, there have been only a few attempts to connect nondeterminism and algebraic specifications. [Subrahmanyam81] and [Broy, Wirsing81] should be mentioned here, which show essentially how to simulate nondeterministic structures by (relatively complex) algebraic specifications of the classical type.

Within the last years there have been attempts to integrate nondeterminism as a basic concept into algebraic specifications ([Nipkow86], [Hesselink88]). These approaches consider operations of algebras as relations, i. e. as set-valued. The notion of a so-called *multi-algebra* ([Pickett67]) could be used there, as well as first similar ideas in [Kapur 80]. Both papers [Nipkow86] and [Hesselink88] treat nondeterministic algebras and basic relations between algebras, but they exclude the question of a well-suited *specification language*. A nondeterministic specification language is presented in [Kaplan 88], but this approach is based on the classical notion of a model and the classical calculus, extended by "built-in" mechanisms for handling sets of data objects. The paper at hands extends the existing work by giving a specification language for the multi-algebra approach.

In a very recent paper [Meseguer 92], the basic idea of using term rewriting as a general framework for computing, without taking care of an equational interpretation, has been covered in detail. In its motivation, these results are very closely related to the work presented here. However, in [Meseguer 92] the semantics are adjusted in such a way that standard term rewriting is sound and complete, leading to a "call-by-name" approach (which is unsound for multi-algebra semantics). The semantics there are mainly oriented towards an initial algebra approach, using category-theoretic tools. In contrast, here the semantics are given by a loose class of multi-algebras in a classical set-theoretic framework, which induces a "call-by-value"-like interpretation. This principle is carried over to the calculus, leading to a calculus which differs from standard term rewriting on the level of deduction systems. Interestingly, the frequently used implementation by graph rewriting turns out to be adequate for our approach, but not completely adequate for standard term rewriting (see section 5.2, example 5.18)!

Another even more recent approach is [Walicki 92/93], where a rather general calculus is introduced for an algebraic treatment of nondeterminism. This work is partially based on earlier versions of our approach. It defines a specification language as well as a sound and complete calculus. However, the syntactical framework used there is much richer than the simple term-rewriting-style of the calculi presented here. It is shown in [Walicki 92/93] that our approach can be seen as a true subcase within the more general framework. The main distinguishing property of our subcase is that we are interested in a programming oriented style of specification, which keeps close connections with term rewriting and admits a direct application of prototyping tools.

0.3 Basic Notions

This section introduces some technical notations which will be used within this book frequently. It may be convenient to skip this section on first reading.

In order to deal with set-valued functions, it is often necessary to construct the *power set* of a given set. The following notation will be used (M is an arbitrary set):

$$\wp(M) = \{N \mid N \subseteq M\}$$
$$\wp^+(M) = \{N \mid N \subseteq M \wedge N \neq \emptyset\}$$
$$\wp_{fin}(M) = \{N \mid N \subseteq M \wedge N \text{ finite}\}$$

Another concept from set theory is the comparison of two arbitrary sets (finite or infinite) with respect to cardinality:

$$|M| \geq |N| \Leftrightarrow_{def} \exists f: M \rightarrow N \text{ and } f \text{ is surjective.}$$

Similarly, sometimes the set of finite sequences over an arbitrary set M is needed, which is denoted by N*. The empty sequence is written as ε, a non-empty sequence is given as a list of its elements, enclosed within angle brackets (⟨⟩). The sequence concatenation operator is an infix operator •, which is defined inductively by the following equations (where s, s' ∈ M*, e ∈ M):

$$\varepsilon \bullet s = s,$$
$$(\langle e \rangle \bullet s) \bullet s' = \langle e \rangle \bullet (s \bullet s').$$

All other notions are common either in the field of algebraic specification or term rewriting. The used notation is similar to [Wirsingetal.83] and [Huet, Oppen80], respectively.

Definition 0.1 (Signature)

> A *signature* is a tuple Σ = (S, F), where S is a set of *sort symbols* and F is a set of *function symbols*. Every function symbol f∈F has a fixed finite sequence of sort symbols (its *argument sorts*) and a sort symbol (its *result sort*).
> The notation [f: $s_1 \times \ldots \times s_n \rightarrow s$]∈F is used to denote a function symbol f∈F with argument sorts $s_1, \ldots, s_n$ and result sort s (s_i, s∈S).
>

The symbol X always means a given countably infinite set of *variable symbols*, where again each x∈X has a fixed sort. More precisely, X is a family of sets of variable symbols:

$$X = (X_s)_{s \in S}.$$

Definition 0.2 (Term)

Let Σ be a signature, X a variable set as above. The set $W(\Sigma, X)_s$ of the Σ, X-*terms* of sort s is the smallest set which fulfils the following conditions:

- Every $x \in X_s$ is contained in $W(\Sigma, X)_s$

- If $[f: s_1 \times \ldots \times s_n \to s] \in F$ and t_i is contained in $W(\Sigma, X)_{s_i}$ (for $1 \le i \le n$), then $f(t_1, \ldots, t_n)$ is contained in $W(\Sigma, X)_s$.

The set $W(\Sigma, \emptyset)_s$ of the *ground terms* of sort s is denoted by $W(\Sigma)_s$. If the sort index of a set of terms $(_s)$ is obvious from the context, it is omitted frequently.   ◊

For the sake of simplicity, all signatures Σ have to be *sensible* as defined in [Huet, Oppen 80], that is for every sort there has to exist at least one ground term.

Definition 0.3 (Subterm, Occurrence)

The mapping Occ computes the set of *occurrences* (or tree addresses) within a term. It is standard to describe such occurrences by finite sequences of natural numbers:

$$\text{Occ: } W(\Sigma, X) \to \wp^+(\mathbb{N}^*)$$

Occ is defined recursively by:

$$\text{Occ}[x] = \{ \varepsilon \} \qquad\qquad \text{if } x \in X_s$$
$$\text{Occ}[f(t_1, \ldots, t_n)] = \{\varepsilon\} \cup \{ i \bullet u \mid i \in \{1, \ldots, n\} \wedge u \in \text{Occ}[t_i] \}$$
$$\text{if } [f: s_1 \times \ldots \times s_n \to s] \in F, \, t_i \in W(\Sigma, X)_{s_i}.$$

t/u denotes the *subterm* of a given term t at the occurrence $u \in \text{Occ}[t]$:

$$t / \varepsilon = t$$
$$f(t_1, \ldots, t_n) / i \bullet u = t_i / u$$

$t[u \leftarrow t']$ denotes the term which results from replacing within t the subterm t/u $(u \in \text{Occ}[t])$ by the term t':

$$t[\varepsilon \leftarrow t'] = t'$$
$$f(t_1, \ldots, t_n)[i \bullet u \leftarrow t'] = f(t_1, \ldots, t_i[u \leftarrow t'], \ldots t_n) \qquad\qquad ◊$$

Vars[t] denotes the set of all variables occurring within a term t:

Vars[t] = { x ∈ X | ∃ u ∈ Occ[t]: t/u = x }

Definition 0.4 (Substitution)

A *substitution* σ is a family of mappings σ = $(\sigma_s)_{s \in S}$ where

$$\sigma_s: X_s \rightarrow W(\Sigma, X)_s$$

such that only for a finite number of x∈X, σ is different from the identity (σ(x) ≠ x). Again, the sort index $(_s)$ is omitted frequently.

A substitution can be easily extended to an endomorphism on W(Σ, X):

$$\sigma(f(t_1, \dots, t_n)) = f(\sigma t_1, \dots, \sigma t_n)$$

The domain of a substitution σ is denoted by

Dom[σ] = { x∈X | σx ≠ x }.

The set of all variables occurring within the substitution terms is denoted by Vars[σ]:

$$Vars[\sigma] = Vars[t_1] \cup \dots \cup Vars[t_n],$$

where $\{t_1, \dots, t_n\}$ = { σx | σx ≠ x }.

A substitution ρ is called a *renaming*, iff ρ is injective and

∀x∈X: ρx∈X.

For two substitutions σ and τ, a composed substitution στ is given by the usual functional composition. The union σ∪τ of two substitutions σ and τ is only defined, if Dom(σ)∩Dom(τ) = ∅; it means to combine σ and τ into a substitution with the domain Dom(σ)∪Dom(τ).

SUBST(Σ, X) is the set of all substitutions σ: X → W(Σ, X), SUBST(Σ) is the set of all *ground substitutions* σ: X → W(Σ).   ◊

A substitution σ, which replaces x∈X by the term t1 and y∈X by the term t2 (and nothing else), is denoted in an explicit notation by: σ = [t1/x, t2/y]. ι is the identity substitution (i.e. ∀x∈X: ι(x) = x).

Given two terms t1 and t2, a substitution σ is called a *unifier* of terms t1 and t2 iff σ t1 = σ t2. If t1 and t2 are unifiable, there is always a *most general unifier* (*mgu*) μ. This means that for every unifier σ for t1 and t2, there is a substitution

λ such that $\sigma = \lambda\mu$. The most general unifier of two terms can be computed efficiently (see for instance [Corbin, Bidoit 83]).

The following sketch of the theory of (equational) algebraic specifications has the only purpose to introduce the notation, for details see [Wirsingetal.83].

A *specification* $T = (\Sigma, E)$ is a tuple, where Σ is a signature and E is a set of *equations* between Σ, X-terms (of the same sort). The central notion for the semantics of such a specification is the notion of a Σ-algebra:

Definition 0.5 (Σ-Algebra)

Let $\Sigma = (S, F)$ be a signature. A Σ-*Algebra* is a tuple $A = (S^A, F^A)$, which consists of:

*   a family of non-empty carrier sets
    $$S^A = (s^A)_{s \in S}, s^A \neq \emptyset \quad \text{for } s \in S$$

*   a family of functions:
    $$F^A = (f^A)_{f \in F}$$
    such that for $[f: s_1 \times \ldots \times s_n \to s] \in F$:
    $$f^A: s_1{}^A \times \ldots \times s_n{}^A \to s^A.$$

The class of all Σ-algebras is called $\text{Alg}(\Sigma)$.   ◊

Within a Σ-algebra A, now the *interpretation* of a term t can be defined. For the interpretation of a non-ground term t, all variables from X occurring in t must be bound to values in A. This is done by a *valuation* β:
$$\beta = (\beta_s)_{s \in S}, \qquad\qquad \beta_s: X_s \to s^A.$$
The interpretation
$$I_\beta^A = (I_{\beta,s}^A)_{s \in S}, \qquad\qquad I_{\beta,s}^A: W(\Sigma, X)_s \to s^A$$

can be defined easily as an extension of the algebra operations. An equation ‹t1 = t2› is called *valid* in A (A |= t1 = t2), iff for all valuations β holds:
$$I_\beta^A[t1] = I_\beta^A[t2].$$

The Σ-algebra A is called a *model* of the specification $T = (\Sigma, E)$, iff all equations in E are valid in A. EqMod(T) denotes the class of all models of the equational specification T.

The calculus of *equational logic* explains how new equations can be deduced from the equations in E. It can be seen as a definition for the following relation on terms:

$$t1 =_E t2 \Leftrightarrow_{def} \exists\ u \in Occ[t1],\ \sigma \in SUBST(\Sigma, X),\ \triangleleft l = r \triangleright \in E:$$
$$t1/u = \sigma l \wedge t2 = t1[u \leftarrow \sigma r]$$

By $=_E^*$ we denote the reflexive-transitive-symmetric closure of $=_E$.

The most important result for equations and equational logic as a specification framework is

Birkhoff's theorem:
$$t1 =_E^* t2 \Leftrightarrow EqMod(T) \models (\ t1 = t2\ ) \qquad\qquad (if\ T = (\Sigma, E))$$

According to this theorem, it is ensured that the calculus can be used only to derive equations which hold in all models of the specification (*soundness*). Moreover, an equation which is valid in all models is deducible with the calculus (*completeness*).

Chapter 1

Nondeterministic Algebraic Specifications

This chapter will show precisely how to generalize the model classes and the specification language for algebraic specifications to the case of nondeterminism. Particular emphasis is laid on a motivation for the design decisions and on a comparison to other approaches.

1.1 Nondeterministic Algebras

An algorithm is called *nondeterministic*, if there are computation states of the algorithm, where the further computation steps are not determined, i. e. where a free choice between different alternatives is admitted. If the final result of the computation is fixed, indepently of the choices, the result is called *determinate*. Here we will study the more general case of nondetermism where even the final result is *non-determinate*. This means that the algorithm may deliver different results when started under equal environment conditions. In a more abstract view, the result of the algorithm is a *set* of possible results (called "breadth" in [CIP 85]).

Nondeterministic programs have been considered already in rather early papers ([McCarthy61], [Floyd 67], [Manna70], [Dijkstra76]). Here the main motivations were:

- The programmer should be freed of unnecessary details at design time ([Dijkstra76]). The design should fix *what* is the function of the program; if there are different ways *how* to realize this function in detail, the decision between them can and should remain open. (A typical example for such a single step with a non-determinate result is: "Choose an arbitrary number between 0 and N".)

- Nondeterminism often is an adequate form of description for a system which depends on unknown parameters. A typical example is an operating system, the behaviour of which depends on the number of users, on the activity of ressources etc. If all these parameters were known, the behaviour of the system would be deterministic. But it is realistic and useful to treat the system without knowing all parameters, consequently to deal with a nondeterministic algorithm ([Hennessy80]).

Both arguments use nondeterminism as a *means of abstraction* for the description of complex systems. This results in a good motivation for integrating nondeterminism into an abstract specification language for the description of algorithms.

For models of algebraic specifications, nondeterminism means that the result of the interpretation of a given function, applied to a given argument, is not fixed uniquely. Below a number of alternative approaches are discussed which try to model this situation mathematically.

1.1.1 A Discussion of Alternative Approaches

Let [f: s → s'] be a function symbol, s and s' sorts of a given signature.

A first variant of nondeterminism is already present within classical algebraic specifications:

(a) Nondeterminism on model level

Let A1 and A2 be two different models of a given specification where:
$$f^{A1}: {}_sA1 \to {}_{s'}A1,$$
$$f^{A2}: {}_sA2 \to {}_{s'}A2,$$
Then for $e \in {}_sA1 \cap {}_sA2$ we may have:
$$f^{A1}(e) = e1, f^{A2}(e) = e2 \text{ and } e1 \neq e2.$$

This example presupposes a so-called *loose semantics* which has been proposed e. g. by [Bauer, Wössner 81], [Wirsing et al. 83]: As the semantics of a specification, a class of models is taken. The result of an operation is not fixed uniquely, since it may differ in different models.

This form of nondeterminism is useful for the description of early phases of a design where design decisions shall be kept open [McCarthy 61]. Within a single model, however, all computations are deterministic.

But sometimes explicitly non-determinate (and therefore nondeterministic) computations are to be described. Abstract specification of programs on operating system level leads to such descriptions, as in the theory of communicating processes. Here the approach described above is no longer adequate, a notion of a model is needed, which admits nondeterministic computations within a single model.

A first option to achieve this aim is the interpretation of a function symbol by a *set of functions*:

(b) Nondeterminism on operation level

Let B be a model of a given specification:
$$f^B = \{ f1, f2 \},$$
$$f1: {}_sB \to {}_{s'}B,$$
$$f2: {}_sB \to {}_{s'}B.$$
Then for $e \in {}_sB$ we may have:
$$f1^B(e) = e1, f2^B(e) = e2 \text{ where } e1 \neq e2.$$

This approach describes precisely the concept of (local) nondeterminism within a functional computation. When a function is applied to given arguments, one out of several prescriptions is chosen to compute the resulting value.

The theory of algebraic specifications stresses the function *application* as the most important operation on functions. It only considers the input-output behaviour of a function. Therefore, an abstraction of the approach (b) also provides an appropriate notion for a model, which uses *set-valued functions*.

(c) Nondeterminism on result level

Let C be a model of a given specification:
$$f^C\colon s^C \to \wp(s'^C).$$
Then for $e \in s^C$ we may have:
$$f^C(e) = \{\, e1, e2 \,\} \text{ where } e1 \neq e2.$$

It is obvious, how for an algebra B corresponding to approach (b) an algebra C corresponding to (c) can be found: Define
$$f^C(e) = \{\, g(e) \mid g \in f^B \,\}$$

This is a true abstraction, i. e. algebras corresponding to (c) contain less information about the structure than in approach (b). Consider the following example:

Let B1 and B2 be algebras according to (b) where:

$$s^{B1} = s'^{B1} = s^{B2} = s'^{B2} = \{\, O, L \,\},$$
$$f^{B1} = \{\, not, id \,\}, f^{B2} = \{\, true, false \,\},$$
$$not, id, true, false\colon \{\, O, L \,\} \to \{\, O, L \,\},$$

$$not(O) = L, \quad not(L) = O,$$
$$id(O) = O, \quad id(L) = L,$$
$$true(O) = L, \quad true(L) = L,$$
$$false(O) = O, \quad false(L) = O.$$

The following algebra C is an abstraction of B1 as well as of B2:

$$s^C = s'^C = \{\, O, L \,\},$$
$$f^C\colon \{\, O, L \,\} \to \wp(\{\, O, L \,\}),$$
$$f^C(O) = \{\, O, L \,\}, \qquad f^C(L) = \{\, O, L \,\}.$$

If we restrict our attention to the input-output behaviour of functions, the function f has the same behaviour in B1 and B2, The function f, applied to an element out of the set { O, L } delivers nondeterministically either O or L.

As long as functions are not considered as objects (like in "higher order" specifications), (c) is equivalent to (b). Since (c) fits well to the abstract style found in algebraic specifications, (c) seems to be better suited for the definition of nondeterministic models. It is interesting to note that exactly the generalization of Σ-algebras described as (c) has already been studied in the Sixties under the name of a *multi-algebra*. As an early source, confer [Pickett67], where for the origin of the notion "multi-algebra" P. Brunovsky (1958) is referred. Below, only the multi-algebra approach will be followed, which forms the basis for the work of [Nipkow86] and [Hesselink88], too. But two other possibilities for introducing nondeterminism should be mentioned before.

(d) Nondeterminism on the level of sorts

Let D be a model of the given specification:
$$f^D\colon \wp(s^D) \to \wp(s'^D).$$
Then for $e \in s^D$ we may have:
$$f^D(\{ e \}) = \{ e1, e2 \} \text{ where } e1 \neq e2.$$

This approach arises as a generalization of (c), by switching from set-valued functions to functions operating on sets. Nevertheless, the specific properties of nondeterministic operations are lost: Simple heterogeneous algebras are considered here, with powersets as its carriers, even non-additive and non-monotone operations are admitted. In [Kaplan 88] this approach is chosen for the description of nondeterminism, but some additional restrictions (in particular the ∪-distributivity af all functions) essentially lead back to the power of the approach numbered (c) here.

A completely different approach, finally, is characterized by a simulation of nondeterminism by deterministic operations:

(e) Nondeterminism by deterministic predicates

Let E be a model of the given specification. Let E contain relations instead of functions:

$$f^E \subseteq s^E \times s'^E$$

The well-known relational product then defines a structure which is comparable to approach (c).

On the model level, it is a matter of taste, whether a functional or relational description is preferred. For instance in [Nipkow86] a relational description for multi-algebras is used; however, an appropriate specification language is not dealt with there. In [Subrahmanyam81], axioms containing nondeterministic operations are translated into axioms for the corresponding predicates; however, the direct relationship between terms and values (interpretation) is lost there. It could be considered an advantage of relational specifications that Prolog-like Horn clauses, if chosen as a specification language, may admit Prolog-like resolution calculi. This idea is investigated in more depth below in chapter 4.

Relationally described nondeterministic specifications obviously are an equivalent, interesting alternative. But for the purposes followed here, this approach has too few similarities with the functional viewpoint of algebraic specifications. In particular, we are interested here in a formal framework which explicitly shows the *principle of uni-directionality* (for instance from input to output) which is central to most programming paradigms. This is the reason why we prefer here the set-valued approach listed above as (c).

Below we will define calculi which correspond directly to specifications of approach (c), and which enable, by term rewriting, a syntactical simulation of computations, too. In this case the uni-directional evaluation by term rewriting corresponds to a *choice* out of a set of possibilities.

1.1.2 The Principle of Extensionality

The *principle of extensionality* is a basic paradigm for applicative and data-flow oriented programming. It means: The identity of a function is determined by its input-output behaviour. Functions with the same input-output-behaviour are considered as equal. This way, we can abstract from the concrete realization, how the function value is computed. A function becomes a "black box" which is observed from the outside only. This point of view has advantages for the modular construction of large systems ("information hiding"). This section will try to motivate the multi-algebra approach again, from the input-output point of view.

In the case of nondeterminism, the interpretation of a function symbol
$$f: s \rightarrow s'$$
in an algebra A can be seen as a computation unit with input and output channels:

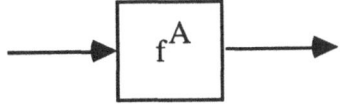

In order to keep with the modularity paradigm, we presuppose that nondeterministic decisions are made *locally* .

In a given nondeterministic computation there is only one value on the input channel. Nondeterministic decisions are made within the computation unit and thus deliver just one value on the output channel.

Experiments with the "black box" f^A consequently may lead only to observations of the shape:

"If x is an input value, y_1, ..., y_n are possible output values."

Approach (c) of the section above exactly mirrors this kind of input-output behaviour.

Approach (b), transferred to our visualisation, would admit additional observations about the way of computation which is chosen by the unit:

"If f^A chooses computation description f_i, then the input value x delivers the output value y."

Approach (d), in contrast, assumes a computation unit f^A, which takes a set of values as its input, and which delivers a set of output values, dependent on the input set:

"If the possible input values are $x_1, ..., x_n$, the possible output values are $y_1, ..., y_m$."

Note that the preference for approach (c) (instead of (b) or (d)) can be motivated only by pragmatic arguments, what is seen as a "realistic" or "interesting" notion for the input-output behaviour.

As a consequence of the choice of approach (c), it is clear now how to define the composition of functions: Just take the input-output behaviour of $f \cdot g$, i. e.:

Given an input value x, possible results of the entire system are exactly the possible results of g^A under input y, where y is a possible output of f^A under x:

$$(f^A \cdot g^A)(e) = \{ e2 \mid e2 \in g^A(e1) \wedge e1 \in f^A(e) \}$$

This means, the *additive* extension of f^A to sets is used. This choice corresponds to the classical relational product and to the usual definitions for multi-algebras.

If these design decisions are compared with those of [Meseguer 92], it is obvious that Meseguer puts more emphasis on fixing the deduction calculus to classical term rewriting. The semantic constructions are adjusted to fit this calculus, whereas here the priority has been set the other way round: We fix the semantic model first, and then adapt the calculus. In fact, the semantics in [Meseguer 92], following an initial algebra approach, are defined in terms of deductions within the rewriting calculus. Further below, the technical differences will be stated more precisely.

1.1.3 The Notion of an Algebra

Using the arguments above, we fix the following generalization of the notion of a heterogeneous Σ-algebra:

Definition 1.1 (Total Σ-Multi-Algebra)

Let $\Sigma = (S, F)$ be a signature. A *(total) Σ-multi-algebra* is a tuple $A = (S^A, F^A)$, which consists of

- a family of non-empty carrier sets
  $$S^A = (s^A)_{s \in S}, s^A \neq \emptyset \text{ for } s \in S$$

a family of set-valued functions where the result is always non-empty:

$$F^A = (f^A)_{f \in F}$$

such that for $[f: s_1 \times \dots \times s_n \to s] \in F$:

$$f^A: s_1{}^A \times \dots \times s_n{}^A \to \wp^+(s^A).$$

The class of all Σ-multi-algebras is denoted by $\text{MAlg}(\Sigma)$.　　　　◊

Below we also use the word "algebra" for a multi-algebra, where no confusion is possible.

It is not completely obvious, why the functions are restricted to deliver only non-empty result sets. For instance in [Hansoul 83], also empty result sets are admitted in multi-algebras. The main argument for exclusion of the empty set is that it somehow represents the *non-existence of a result*, which is equivalent to undefinedness. It is well known from the theory of algebraic specifications that an adequate integration of partial functions leads to a number of serious problems. So the question of partiality is postponed to chapter 6, and functions are restricted here to "total" ones, which always deliver at least one result.

Example 1.2

Let $\Sigma = (S, F)$ be the following signature:

sort　Nat

func　zero: $\to$ Nat,　　　　　succ: Nat $\to$ Nat,

　　　　add: Nat $\times$ Nat $\to$ Nat,　　or: Nat $\times$ Nat $\to$ Nat,

　　　　some: $\to$ Nat

The algebra A then is a Σ-multi-algebra, where

$$\text{Nat}^A = \mathbb{N},$$

$\text{zero}^A: \to \wp^+(\mathbb{N}),$　　　$\text{zero}^A = \{\, 0 \,\},$

$\text{succ}^A: \mathbb{N} \to \wp^+(\mathbb{N}),$　　$\text{succ}^A(n) = \{\, n+1 \,\},$

$\text{add}^A: \mathbb{N} \times \mathbb{N} \to \wp^+(\mathbb{N}),$　$\text{add}^A(n,m) = \{\, n+m \,\},$

$\text{or}^A: \mathbb{N} \times \mathbb{N} \to \wp^+(\mathbb{N}),$　$\text{or}^A(n,m) = \{\, n \,, m \,\},$

$\text{some}^A: \to \wp^+(\mathbb{N}),$　　$\text{some}^A = \mathbb{N}$

　　　　　　　　　　　　　(where m, n$\in \mathbb{N}$).

Another Σ-multi-algebra is B, given by

$\text{Nat}^B = \{ Z, N \}$,

$\text{zero}^B = \{ Z \}$, $\text{succ}^B(e) = \{ N \}$,

$\text{add}^B(Z,e) = \{ e \}$, $\text{add}^B(N,e) = \{ N \}$,

$\text{or}^B(e1,e2) = \{ e1, e2 \}$, $\text{some}^B = \{ Z, N \}$

                              (where $e, e1, e2 \in \{ Z, N \}$) $\Diamond$

In the algebra A of this example, the operations or^A and some^A are called (truly) nondeterministic, since they actually do deliver a choice between different results. The other operations are single-valued functions and therefore called deterministic.

The operation some^A in the example above shows that multi-algebras in general admit infinite result sets (indicating a choice out of infinitely many values, which is sometimes called *unbounded* nondeterminism). Please note that the choice of names in the example above shows a close correspondence to the semantics assigned to them by the multi-algebra A. However, the formal specification (which is only a signature up to now) does not resemble these informal ideas at all. It is easy to give a multi-algebra for the same signature where some is a deterministic operation and zero a nondeterministic one!

The set of ground terms can be made into a multi-algebra (since every Σ-algebra is a Σ-multi-algebra with singleton result sets, too). This is a very particular algebra, where every operation is deterministic.

Example 1.3

For an arbitrary signature Σ, a Σ-multi-algebra $W\Sigma$ (the *term algebra*) is given by:

$s^{W\Sigma} = W(\Sigma)_s$ for $s \in \Sigma$

$f^{W\Sigma}(t_1, \ldots, t_n) = \{ f(t_1, \ldots, t_n) \}$ for $f \in F$. $\Diamond$

In order to define the interpretation of terms in a multi-algebra, the operations of the algebra just have to be composed as the structure of the term indicates. However, for giving a meaning also to terms containing variables, we need a notion of an *environment* which binds variables. A term with variables can be interpreted only if the environment defines fixed values for the variables. It is consequent to admit here as values to be assigned to variables only *single values* out of the carrier set (no set-valued environments). The reason for this is that a computation within a given algebra contains only single values as intermediate results. Within the informal description of section 1.2 above, set-valued environments would correspond to observations under a fixed set of possible input values. It is clear that such observations can be composed out of observations with single-valued environments, by additive extension.

Definition 1.4 (Environment)

Let $A = (S^A, F^A)$ be a Σ-multi-algebra.
An *environment* β of X in A is a family
$$\beta = (\beta_s)_{s \in S}$$
of mappings
$$\beta_s: X_s \to s^A.$$
ENV(X, A) denotes the set of all environments of X in A.   ◊

The definition of interpretation composes the operation provided by the algebra, using an additive extension, when applying a function to a set of values:

Definition 1.5 (Interpretation)

Let $A \in MAlg(\Sigma)$, $\beta \in ENV(X, A)$.
The *interpretation* I_β^A is a family of mappings
$$I_\beta^A = (I_{\beta,s}^A)_{s \in S}, \qquad I_{\beta,s}^A: W(\Sigma, X)_s \to \wp^+(s^A) \qquad \text{for } s \in S.$$

$I_{\beta,s}^A$ is defined inductively as follows:

(1) If $t = x$ and $x \in X_s$:
$$I_{\beta,s}^A [t] = \{ \beta(x) \}$$

(2) If $t = f(t_1, ..., t_n)$ where $[f: s_1 \times ... \times s_n \to s] \in F$
     ($t_i \in W(\Sigma, X)_{s_i}$ for all $i \in \{1, ... , n\}$):
$$I_{\beta,s}^A [t] = \{ e \in f^A(e_1,...,e_n) \mid e_i \in I_{\beta,s_i}^A [t_i], 1 \leq i \leq n \}$$

If t is a ground term (out of $W(\Sigma)_s$), we write also I^A instead of $I_{\beta,s}^A$. ◊

Example 1.6

In the multi-algebra A from example 1.2 the following propositions hold:

$$I^A[\text{ zero }] = \{\ 0\ \}$$
$$I^A[\text{ add(or(zero,succ(zero)),or(zero,succ(zero))) }] = \{\ 0\ ,1\ ,2\ \}$$

If β is an environment with $\beta(x) = 0$, $\beta(y) = 1$, then we have:

$$I^A_\beta[\text{ add(x,or(x,y)) }] = \{\ 0\ ,1\ \}.$$

In the multi-algebra B the following holds:

$$I^B[\text{ zero }] = \{\ Z\ \}$$
$$I^B[\text{ add(or(zero,succ(zero)),or(zero,succ(zero))) }] = \{\ Z\ ,N\ \} \quad \Diamond$$

1.2 Inclusion Rules as a Specification Language

The semantics of a classical algebraic specification is given by a class of algebras which are characterized by a set of axioms. Basically, algebraic specifications may use arbitrary first-order formulas as axioms ([Wirsing et. al. 83]), where the atomic formulas are equations.

Particular interest has been paid to specifications where the axioms are only equations or conditional equations (*positive conditionals*) with universally quantified variables. On the one hand, the model class always has a nice lattice structure in this case; on the other hand, the equational calculus is particularly simple and therefore well-suited for support by software tools. These tools usually are based on term rewriting (rewriting engine, Knuth-Bendix-completion, E-unification by narrowing).

1.2.1 Axioms and their Semantics

When switching to the nondeterministic case, we first have to find an appropriate notion replacing the equations. The purpose of these atomic formulas is to describe the nondeterministic choice out of several possibilities.

Such a process of decision does not preserve the information contained in a term, but may reduce the amount of information. Therefore, we use unsymmetric (oriented) atomic formulas. Below, "inequations" play the role which equations have in the classical case. An inequation between two terms t1 and t2 is denoted

$$t1 \rightarrow t2.$$

It is to be understood informally as:

> "Every (nondeterministic) possibility for the evaluation of t2 is a (nondeterministic) possibility for the evaluation of t1, too."

With respect to the intended interpretation, we call the inequations from now on *inclusion rules*. The notation for inclusion rules is the same as it is standard for rewrite rules, because below a tight correspondence to term rewriting will be developed.

Definition 1.7 (Inclusion Rule)

> An *(atomic) (Σ, X-)inclusion rule* is a pair of terms of equal sort, which is denoted as a formula
>
> $$t1 \rightarrow t2$$
>
> where $t1, t2 \in W(\Sigma, \Xi)_s, s \in S.$ ◊

Variables occuring in inclusion rules are implicitely understood as universally quantified (like in equational specifications).

The validity of an inclusion rule has to take care of the orientation. In contrast to equational logic, which centers around the notion of equality, our objects are decision processes, and this leads to *set inclusion*.

Definition 1.8 (Validity)

> Let A be a Σ-multi-algebra. An inclusion rule $t1 \rightarrow t2$ is called *valid* in A, written:
>
> $$A \models t1 \rightarrow t2$$
>
> iff for all environments $\beta \in ENV(X, A)$:
>
> $$I^A_\beta [t1] \supseteq I^A_\beta [t2].$$ ◊

Note that this definition of validity relies on the definition of an environment: A variable within an axiom always means a (determinate) object, and not a nondeterministic expression.

Example 1.9

Within the multi-algebras A und B of example 1.2 the following inclusion rules are valid:

add(zero,x) $\rightarrow$ x,	add(succ(x),y) $\rightarrow$ succ(add(x,y))
or(x,y) $\rightarrow$ x,	or(x,y) $\rightarrow$ y
some $\rightarrow$ zero,	some $\rightarrow$ succ(some) $\Diamond$

Now we can define nondeterministic algebraic specifications in analogy to the standard approach.

Definition 1.10 (Nondeterministic Algebraic Specification)

A *(nondeterministic) (algebraic) specification* is a tuple T = (Σ, R), which consists of a signature Σ and a finite set R of Σ, X-inclusion rules, which are called the *axioms* of the specification. $\Diamond$

Definition 1.11 (Model)

A nondeterministic Σ-algebra A is called a *model* of the nondeterministic specification T = (Σ, R), iff for all inclusion rules $\Phi \in$ R: A $\models \Phi$. Mod(T) denotes the class of all models of the specification T . $\Diamond$

Example 1.12

This first example of a specification combines the signature from example 1.1 with the inclusion rules from example 1.9. We use a notation similar to many standard specification languages (for instance OBJ, PLUSS):

spec NAT
sort Nat
func zero: $\rightarrow$ Nat,   succ: Nat $\rightarrow$ Nat,
         add: Nat $\times$ Nat $\rightarrow$ Nat,   or: Nat $\times$ Nat $\rightarrow$ Nat

axioms

add(zero,x) $\rightarrow$ x,   add(succ(x),y) $\rightarrow$ succ(add(x,y)),

or(x,y) $\rightarrow$ x,   or(x,y) $\rightarrow$ y

some $\rightarrow$ zero,   some $\rightarrow$ succ(some)

end

The algebra A from example 1.2 now is a model of NAT. $\Diamond$

1.2.2 The Calculus of Term Rewriting

As already indicated above, the axioms of a nondeterministic specification shall be used for a calculus for the derivation of further properties, which hold within the model class of a specification. The tight relationship to the formalism of term rewriting will lead to a situation, where this calculus can be seen as an operational interpretation of the specification as well.

The well-founded theory of term rewriting (see [Huet, Oppen 80]) offers an "oriented" analogon to equational logic (see also [Meseguer 92]). The orientation of the rules admits a much more efficient way to carry out deductions in comparison to equational logic. This is the reason why term rewriting forms the basis for most of the software tools available for equational specifications now (e. g. OBJ, ASSPEGIQUE, AXIS, OBSCURE, RAP).

The classical theory of term rewriting also assumes a finite set R of inclusion rules, but as an additional restriction for all axioms $\langle t1 \rightarrow t2 \rangle \in R$ it is required that the variable condition holds:

Vars[t1] $\supseteq$ Vars[t2].

This requirement can be omitted in our approach, since it aims mainly at the notion of confluence, a condition which is always violated by non-trivial nondeterministic specifications (if considered as a system of rewrite rules). See below for a formal definition of confluence.

The term rewriting relation $\rightarrow_R$ (for a given axiom set R) is a binary relation between terms of the same sort:

t1 $\rightarrow_R$ t2 $\Leftrightarrow$ def

$\exists$ u $\in$ Occ[t1], $\sigma \in$ SUBST(Σ, X), $\langle l \rightarrow r \rangle \in R$:

t1/u = σl $\wedge$ t2 = t1[u $\leftarrow$ σr]

The symbol $\to_R^*$ means the reflexive-transitive closure of $\to_R$.

This notion exactly describes the "oriented" generalization of the relation $=_E$ known from equational logic.

The theory of term rewriting becomes useful for equational logic by means of the property of *confluence*:

A term rewriting system R is called *confluent* iff
$$\forall\ t1, t2, t3: (\ t1 \to_R^* t2 \wedge t1 \to_R^* t3\ )$$
$$\Rightarrow\ \exists\ t4: (\ t2 \to_R^* t4 \wedge t3 \to_R^* t4\ ).$$
In words, the result of every rewriting sequence has to be determined independently of the actual choice of the rewriting steps.

The theorem below connects equational logic with term rewriting:

Confluence Theorem:
> If R is confluent, then:
> $$t1 =_R^* t2 \Leftrightarrow (\ \exists\ t3:\ t1 \to_R^* t3 \wedge t2 \to_R^* t3\ )$$

It is obvious that in the case of nondeterminism the confluence condition will *not* hold: Here we are interested explicitly in deriving several different results for a given term.

Below we will try to circumvent the confluence theorem. We aim at a result similar to Birkhoff's theorem for general, non-confluent term rewriting. Ideally we would try to prove:
$$t1 \to_R^* t2 \Leftrightarrow Mod(R) \models (\ t1 \to t2\ )$$

The proposition above does not hold in general, but it holds under specific preconditions which will be explained later. The driving idea of this and the next chapter will be to isolate particular circumstances under which a usable variant of this proposition does hold.

First we will give an alternative formulation of $\to_R^*$, then the $\Rightarrow$-part of the proposition above (soundness) will be studied, followed by the $\Leftarrow$-part (completeness). This development cycle will be repeated three times until a sufficient solution is reached.

The term rewriting relation $\to_R^*$ can be represented also by a calculus which can be used to derive formulas of the shape $t1 \to t2$ from the given set of inclusion rules. This form of description (which is used consequently for instance in [Padawitz 88]) is easier to handle within proofs.

Definition 1.13 (Calculus of Term Rewriting)

Let $T = (\Sigma, R)$ be a nondeterministic algebraic specification. Then a formula $t1 \to t2$ is called *deducible* in T, written as

$$T \vdash_{RC} t1 \to t2,$$

iff there is a formal derivation for $t1 \to t2$ using the following deduction rules:

(REFL)
$$\frac{}{t \to t} \quad \text{if } t \in W(\Sigma, X)$$

(TRANS)
$$\frac{t1 \to t2 \,, t2 \to t3}{t1 \to t3} \quad \text{if } t1, t2, t3 \in W(\Sigma, X)$$

(CONG)
$$\frac{t_i \to t_i'}{f(t_1, ..., t_{i-1}, t_i, t_{i+1}, ..., t_n) \to f(t_1, ..., t_{i-1}, t_i', t_{i+1}, ..., t_n)}$$

$$\text{if } [f: s_1 \times ... \times s_n \to s] \in F,$$
$$t_j \in W(\Sigma, X)_{s_j} \text{ where } j \in \{1, ..., n\}, t_i' \in W(\Sigma, X)_{s_i}$$

(AXIOM)
$$\frac{}{\sigma l \to \sigma r} \quad \begin{array}{l} \text{if } \langle l \to r \rangle \in R, \sigma \in \text{SUBST}(\Sigma, X) \\ \text{(i.e. } \sigma: X \to W(\Sigma, X)) \end{array} \quad \Diamond$$

The notation $\vdash_{RC}$ (rewriting calculus) has been used in order to distinguish the calculus from similar calculi which will be introduced below. In the following, sometimes $R \vdash_{RC}$ is written which is meant as a synonym for $T \vdash_{RC}$ (T = (Σ, R) a specification). If it is clear from the context which specification is meant, the notation is further simplified by omitting T or R, respectively. (The same notational convention will be applied for all other calculi introduced below.)

1.2.3 Soundness: A Negative Result

When a logical calculus is analysed, the most important (and therefore first) question is whether it is *sound* with respect to the underlying semantics. Soundness means that the calculus allows us only to derive statements which are semantically valid.

It is this question which already leads into problems for the calculus above. But the difficulties appearing here are well-known from the semantics of nondeterministic programming languages. In [Bauer, Wössner 81], for instance, the following example is mentioned.

Let a function declaration (in a classical algorithmic language) be given
    funct double = (**nat** n)**nat**: n + n
which computes for a natural number the double of its numerical value.

It is now interesting to consider a call of this function with a nondeterministic expression as its argument, e. g.
    double(zero_or_one)
where zero_or_one means the nondeterministic choice between the values 0 and 1. Basically, two points of view can be thought of, which are known as "call-time-choice" and "run-time-choice" ([Hennessy 80], [Benson79]).

"Call-time-choice" assumes that the function gets only deterministic *objects*, and not nondeterministic expressions, as its arguments. The call above therefore is equivalent to one of the both calls
    double(0) or double(1) ,
thus the possible results are described by the set { 0, 2 }.

"Run-time-choice", however, treats the call as equivalent to
    zero_or_one + zero_or_one,
which is similar to the "copy rule" of ALGOL 60: The nondeterministic expression is copied into the function body. Now the values out of { 0, 1, 2 } are admissible results. The value 1 is legal, since both "copies" of zero_or_one may choose the result value independently one of the other.

The semantics introduced in the sections 1.1 and 1.2 correspond to call-time-choice. Unfortunately, the term rewriting calculus corresponds to run-time-choice (copy rule), as it can be seen from the following example:

Example 1.14

> **spec** DOUBLE
>
> **sort** Nat
>
> **func** zero: $\to$ Nat,   succ: Nat $\to$ Nat,
>
>   add: Nat × Nat $\to$ Nat,   double: Nat $\to$ Nat,
>
>   zero_or_one: $\to$ Nat
>
> **axioms**
>
>   add(zero,x) $\to$ x,
>
>   add(succ(x),y) $\to$ succ(add(x,y)),
>
>   double(x) $\to$ add(x,x),
>
>   zero_or_one $\to$ zero,
>
>   zero_or_one $\to$ succ(zero)
>
> **end**

A model of DOUBLE is e. g. the algebra N below:

$$\text{Nat}^N = \mathbb{N},$$
$$\text{zero}^N = \{\ 0\ \},\qquad\qquad \text{succ}^N(n) = \{\ n{+}1\ \},$$
$$\text{add}^N(n,m) = \{\ n{+}m\ \},\qquad \text{double}^N(n) = \{\ 2n\ \},$$
$$\text{zero\_or\_one}^N = \{\ 0\ ,1\ \}$$

A derivation within the term rewriting calculus is:

(1) $\vdash_{RC}$ zero_or_one $\to$ zero   (AXIOM)

(2) $\vdash_{RC}$ zero_or_one $\to$ succ(zero)   (AXIOM)

(3) $\vdash_{RC}$ add(zero_or_one,zero_or_one) $\to$ add(zero,zero_or_one)

                                                          (CONG), (1)

(4) $\vdash_{RC}$ add(zero,zero_or_one) $\to$ add(zero,succ(zero))

                                                          (CONG), (2)

(5) $\vdash_{RC}$ add(zero_or_one,zero_or_one) $\to$ add(zero,succ(zero))

                                                          (TRANS), (3), (4)

(6) $\vdash_{RC}$ add(zero,succ(zero)) $\to$ succ(zero)   (AXIOM)

(7) $\vdash_{RC}$ add(zero_or_one,zero_or_one) $\to$ succ(zero)

                                                          (TRANS), (5), (6)

(8) $\vdash_{RC}$ double(zero_or_one) $\to$ add(zero_or_one,zero_or_one)

                                                          (AXIOM)

(9) $\vdash_{RC}$ double(zero_or_one) $\to$ succ(zero)   (TRANS), (7), (8)

But in N the following inclusion does not hold:

double(zero_or_one) $\rightarrow$ succ(zero)

(since I^N[double(zero_or_one)] = { 0 , 2 }). $\Diamond$

The example shows that (AXIOM) does not treat the inclusion rule

double(x) $\rightarrow$ add(x,x)

in a sound way, because both "copies" of the term which is substituted for x (zero_or_one in the example), can be evaluated independently.

It must be decided now whether the semantic concepts from above (in particular the interpretation of a term) should be revised, or whether the calculus should be modified. There are good reasons why the given semantical framework has been chosen. The basic assumption that a variable always stands for a single value comes from the intention to model a somehow "realistic" scenario for nondeterministic computation, where only single data items are transmitted between computational units. Therefore the deductions have to to be adapted to the semantic framework.

It is an interesting observation that the soundness problem does not appear in [Meseguer 92]. There the semantics are adjusted in such a way that the rewriting calculus RC is sound and complete. This excludes models like N above, but it contradicts to our basic paradigm that a function gets only single values as its input, and therefore variables always stand for single values. To put it simply, [Meseguer 92] uses a "run-time-choice" strategy, in difference to the "call-time-choice" which is preferred here.

There are various possibilities to refine the calculus in such a way that it becomes sound with respect to the multi-algebra semantics studied here. Basically, two approaches are most promising:

(1)   The rewriting calculus can be changed in such a way that it truly reflects a "call-time-choice" strategy. For this purpose there must be some syntactical possibility to recognize whether a term is determinate, that is whether it is always interpreted by a singleton set.

(2)   The application of an axiom can be adapted in such a way that it keeps the information about which term is a physical copy of another one. This is a step towards rewriting on graph-like structures.

Both approaches will be worked out in more detail in later chapters (approach (1) in chapter 2, approach (2) in chapter 5, section 5.2). Another example is very helpful for deciding which way to go for the adaptation of the calculus now.

Example 1.15

The algebra NN below is another model of the specification DOUBLE defined in the preceding example 1.14:

$Nat^{NN} = \mathbb{N}$
$zero^{NN} = \{ 0 , 1 \},$ $\qquad\qquad$ $succ^{NN}(n) = \{ n+1 \},$
$add^{NN}(n,m) = \{ n+m \},$ $\qquad$ $double^{NN}(n) = \{ 2n \},$
$zero\_or\_one^{NN} = \{ 0 , 1 , 2 \}$

The model NN shows that even the most simple application of the axiom for double is unsound. In NN we have:

$$\neg\ (NN \models double(zero) \twoheadrightarrow zero),$$

since

$$I^{NN}[double(zero)] = \{ 0 , 2 \}, \qquad I^{NN}[zero] = \{ 0 , 1 \}.$$

So any calculus which allows us to deduce
$$double(zero) \twoheadrightarrow zero$$
is not sound.   ◊

This example shows that there is a rather general problem in treating inclusion rules with multiple variable occurrences on the right hand side. In fact it even shows that the expressivity of the specification language is still too low, since intuitively we would expect the inclusion $double(zero) \twoheadrightarrow zero$ to hold in all models of DOUBLE. But this expectation implicitly uses the assumption that zero is a deterministic function, which is not the case in this counterexample.

Considering approach (2), it does not suffice to extend the term rewriting calculus by a notion of rewriting on terms with sharing of subterms. We use the notation
$$\textbf{let } x = zero \textbf{ in } add(x,x)$$
to denote a term which contains two shared occurrences of the subterm zero. It is not difficult to extend the interpretation to such terms with sharing in a way which ensures that both copies of the shared term (zero) always are evaluated to the same elementary value. So the interpretation of the **let**-term from above in

NN is intended to be $\{0, 2\}$. The calculus can be extended in such a way that it allows us to derive the inclusions

     |- double(zero) $\rightarrow$ **let** x = zero **in** add(x,x)

     |- add(zero,zero) $\rightarrow$ zero.

However, the inclusion **let** x = zero **in** add(x,x) $\rightarrow$ add(zero,zero) again does not hold in NN (the first term has the interpretation $\{0, 2\}$, the second one means $\{0, 1, 2\}$.) So also a sound graph-rewriting calculus should not admit the application of the add-rule to the term containing the shared subterm zero. Unfortunately, standard graph rewriting, as it is defined for instance in [Barendregt et al. 87], would perfectly admit the application of the add-rule. See chapter 5 (section 5.2) for a re-examination of this idea in a more detailed framework.

However, approach (2) can be dealt with nicely within the new framework of [Walicki 92/93]. This work leads to a richer syntax, where deductions are made in a context consisting of *variable bindings*. We use here the notation "x∈t" to express that x is bound to some value out of the interpretation of term t; and implication to prefix an inclusion with such a binding context. From this notation, it should be obvious how the semantics can be extended. Within such a calculus, we can derive the formula

     $x \in$ zero $\Rightarrow$ double(x) $\rightarrow$ add(x,x),

which is the most refined statement about double(zero) which can be deduced soundly. However, this approach uses the deduction of conditional statements which is a significant step beyond standard (and even conditional) term-rewriting.

The next chapter addresses an extension of the specification language following the approach number (1). On a first reading, it is recommended to skip directly to this chapter 2 from here. The section 1.2.4, which follows immediately below, just studies an interesting special case for which classical term rewriting is sound and complete. Unfortunately, this special case excludes almost all realistic software specifications, so it is interesting only from the theoretical point of view.

1.2.4 Right-Linearity: A Special Case

It is quite obvious that all the difficulties discussed in the section above came from axioms which contained multiple occurrences of a variable within their right hand sides. The idea of this section is to exclude such multiple occurrences syntactically. It turns out that for this special case a general soundness and completeness result holds.

A term which contains exactly one occurrence for every one of its variables is called *linear*. The next definition carries this definition over to systems of inclusion rules.

Definition 1.16 (Linearity)

A term $t \in W(\Sigma, X)$ is called *linear* iff there are no multiple occurrences of a variable within it, i. e.:
$$\forall x \in X: \forall u1, u2 \in Occ(t): (t/u1 = x) \wedge (t/u2 = x) \Rightarrow (u1 = u2).$$

An inclusion rule $\langle l \rightarrow r \rangle$ is called *right-linear* iff the term r is linear.
A set R of inclusion rules is called right-linear iff all axioms in in R are right-linear. ◊

The following theorem shows that a restriction to right-linearity entails soundness of classical term rewriting under nondeterministic interpretations.

Theorem 1.17 (Soundness)

Let $T = (\Sigma, R)$ be a nondeterministic algebraic specification where R is right-linear. Then for t1, t2 $\in W(\Sigma, X)$ holds:
$$T \vdash_{RC} t1 \rightarrow t2 \Rightarrow Mod(T) \models t1 \rightarrow t2 .$$

Proof:

The proof of soundness is done by induction on the (length of the) derivation. When in this proof the deduction rule (AXIOM) is considered, the condition of right-linearity is necessary for the application of the following lemma:

Lemma 1.17.1

Let $A \in Mod(T)$, $\beta \in ENV(X, A)$, $\sigma \in SUBST(\Sigma, X)$.
Then for $t \in W(\Sigma, X)$ holds: (*)
$$I^A_\beta[\sigma t] \supseteq \{e \in I^A_\gamma[t] \mid \gamma \in ENV(X, A) \wedge \forall x \in Vars[t]: \gamma x \in I^A_\beta[\sigma x]\}$$

If t is linear, within the proposition (*) set equality holds.

The proof of lemma 1.17.1 is given in appendix A. ◊

In order to prove also the completeness of term rewriting, a term model is constructed now, similar to classical equational logic. However, the construction follows the idea of an ideal completion (confer [Möller 82]) instead of forming a quotient of the set of terms.

Definition 1.18 (Term Algebra WΣ/R)

For a given signature Σ and a set of axioms R, a Σ-Algebra WΣ/R is constructed by:

$$s^{W\Sigma/R} = W(\Sigma, \Xi)_s \qquad \text{for } s \in S$$
$$f^{W\Sigma/R}(t_1, \ldots, t_n) = \{\ t \in W(\Sigma, X) \mid R \vdash_{RC} f(t_1, \ldots, t_n) \to t\ \}$$
$$\text{for } f \in F.$$

$s^{W\Sigma/R} \neq \emptyset$ holds, if for all sorts there is at least one variable. According to (REFL) then $f^{W\Sigma/R}(t_1, \ldots, t_n) \neq \emptyset$.   ◊

Theorem 1.19

Let $T = (\Sigma, R)$ be a specification where $l \in X$ for all $\langle l \to r \rangle \in R$. Then WΣ/R is a model of T.

Proof: See appendix A.   ◊

Example 1.20

Within the term model WΣ/DOUBLE for example 1.14 we have:

$I^{W\Sigma/DOUBLE}[double(zero\_or\_one)] =$
      $\{$ double(zero_or_one) , add(zero_or_one,zero_or_one) ,
      add(zero_or_one, zero) , add(zero,zero_or_one),
      add(zero,succ(zero)) , add(succ(zero),zero),
      add(succ(zero),succ(zero)) , succ(add(zero,zero)),
      succ(add(zero,succ(zero))), succ(succ(add(zero,zero))),
      succ(succ(zero)), succ(zero), zero $\}$,

$I^{W\Sigma/DOUBLE}[succ(zero)] = \{$ succ(zero) $\}$,
i. e.:   WΣ/DOUBLE $\models$ double(zero_or_one) $\to$ succ(zero)   ◊

A small example may illustrate why the precondition

$l \notin X$ für $\langle l \to r \rangle \in R$

is necessary to ensure that the term model in fact is a model of the specification.

Example 1.21

> **spec** LD
> **sort**   s
> **func**   a: $\to$ s, b: $\to$ s
> **axioms**
>
>   x $\to$ a
> **end**

The algebra WΣ/LD is not a model of LD, as can easily be seen. Let β be an environment assigning a term to the variable x. Then according to definition 1.5:

$$I_\beta^{W\Sigma/R}[x] = \{ \beta x \}.$$

If the environment β is specialized to assign the term b to the variable x, this means:

$$I_\beta^{W\Sigma/R}[x] = \{ \beta \} \not\supseteq \{ a \} = I_\beta^{W\Sigma/R}[a],$$

so the single inclusion rule of LD does not hold in WΣ/LD.   ◊

Example 1.21 also illustrates that the condition "$l \notin X$" (which is sometimes called *left-definiteness*) is a necessary prerequisite for completeness. Without it completeness is lost.

Example 1.22

Consider again the specification LD from example 1.21.

The axiom x $\to$ a forces the interpretation of a and b to be equal within all models of LD :
Let A be a model of LD , $e_b \in I^A[b]$.
Using the environment $\beta(x) = e_b$, the axiom has to hold, thus:

$$\{ e_b \} \supseteq I^A[a].$$

This means (because of $I^A[a] \neq \emptyset$):

$$e_b \in I^A[a].$$

Therefore holds:

$$I^A[a] \supseteq I^A[b].$$

Thus: $Mod(LD) \models a \to b$, although this inclusion is not deducible. $\Diamond$

With the appropriate preconditions, however, there is a completeness result. Please remember that this completeness does only make sense for right-linear specifications, because this is the case for which the soundness of the term rewriting calculus RC has been proven.

Theorem 1.23 (Completeness)

Let $T = (\Sigma, R)$ be a nondeterministic algebraic specification where $1 \notin X$ holds for all $\langle l \to r \rangle \in R$. Then for $t1, t2 \in W(\Sigma, X)$ holds:

$$Mod(T) \models t1 \to t2 \quad\quad \Rightarrow \quad T \vdash_{RC} t1 \to t2.$$

Proof:

Completeness follows from the existence of the term model $W\Sigma/R$. (For the lemma 1.19.1, see appendix A.)

$$\begin{array}{ll} Mod(T) \models t1 \to t2 \quad \Rightarrow & \text{(Thm. 1.16)} \\ W\Sigma/R \models t1 \to t2 \quad \Rightarrow & \text{(Defn. 1.8)} \\ I_l^{W\Sigma/R}[t1] \supseteq I_l^{W\Sigma/R}[t2] \Rightarrow & \text{(Lemma 1.19.1)} \end{array}$$

$$\{ t \mid R \vdash_{RC} t1 \to t \} \supseteq \{ t \mid R \vdash_{RC} t2 \to t \} \Rightarrow$$
$$(\text{since } \vdash t2 \to t2)$$
$$R \vdash_{RC} t1 \to t2 \quad\quad\quad\quad\quad\quad\quad\quad\quad \Diamond$$

Unfortunately, the restriction to right-linear specifications is too strong to be acceptable for a practical specification technique. For instance, the standard description for the multiplication of natural numbers already contains non-right-linear inclusion rules. Example 1.14 above also shows that even quite "natural" specifications violate the right-linearity condition. Chapter 2 therefore discusses ways for the construction of a more general calculus which still remains very similar to term rewriting.

Chapter 2

Specifications with a Deterministic Basis

The conclusions from chapter 1 are:

(1) Classical term rewriting is unsound for nondeterministic specifications.

(2) If the axioms are restricted to right-linear inclusion rules, classical term rewriting is sound and complete, however this restriction is not satisfactory for practical applications.

Moreover, chapter 1 gave indications that the specification language of inclusion rules itself is too simple to designate an appropriate model class. In particular, it does not provide any way to express that some term is deterministic, this is that it must always be interpreted by a single value.

In this chapter, the language is extended by a particular kind of formulae which state explicitly for a term that it must have a one-element interpretation. This restricted language admits a sound calculus, which is very close to classical term rewriting. Under reasonable preconditions, also completeness can be shown.

The essential idea for the refinement is to designate a "basis" part of a specification which is called deterministic, because it must always be interpreted determinately.

2.1 Deterministic Basis

As good starting point for the development of a sound calculus, example 2.1 recalls examples 1.14 und 1.15, which showed that term rewriting is not sound in general.

Example 2.1

The algebra NN below was defined in example 1.15:

$\text{Nat}^{\text{NN}} = \mathbb{N}$

$\text{zero}^{\text{NN}} = \{\ 0\ ,\ 1\ \},$ $\text{succ}^{\text{NN}}(n) = \{\ n{+}1\ \},$

$\text{add}^{\text{NN}}(n,m) = \{\ n{+}m\ \},$ $\text{double}^{\text{NN}}(n) = \{\ 2n\ \},$

$\text{zero\_or\_one}^{\text{NN}} = \{\ 0\ ,\ 1\ ,\ 2\ \}$

The model NN fulfils the axioms from example 1.14:

$\quad\quad$ add(zero,x) $\rightarrow$ x,

$\quad\quad$ add(succ(x),y) $\rightarrow$ succ(add(x,y)),

$\quad\quad$ double(x) $\rightarrow$ add(x,x),

$\quad\quad$ zero_or_one $\rightarrow$ zero,

$\quad\quad$ zero_or_one $\rightarrow$ succ(zero),

but not the inclusions listed below (which are nevertheless deducible by term rewriting):

$\quad\quad$ double(zero_or_one) $\rightarrow$ add(zero_or_one,zero_or_one),

$\quad\quad$ double(zero) $\rightarrow$ add(zero,zero). $\lozenge$

Obviously, the "mistake" comes from the application of the non-right-linear rule. However, with an intuitive idea of the specification in mind, one would expect that at least the inclusion

(*)   double(zero) $\rightarrow$ add(zero,zero)

does hold in all models. This intuitive interpretation always assumes the well-known symbol "zero" to be interpreted as the singleton set $\{\ 0\ \}$. Here the model semantics contradicts intuition.

The other inclusion

(**)   double(zero_or_one) $\rightarrow$ add(zero_or_one,zero_or_one)

is not an intuitive consequence of the axioms (since "zero_or_one" is obviously a nondeterministic function symbol). Here the deduction semantics given by term rewriting is counterintuitive.

To capture this idea, the fact must be formalized that zero is a deterministic operation for all models. This leads to an exclusion of the "non-standard" model NN (where zero is interpreted by a choice between two values). The calculus then must be adapted in such a way that it admits the deduction of (*), but not of (**).

2.1.1 Soundness and Deterministic Basis

The specification language has to be enriched by a means to state whether the result of a function application is determinate or not. So the specification gains a *deterministic basis* part, enriched by possibly nondeterministic extensions. This concept coincides with the basic design decision for our theory which studies nondeterministic functions working on a set of (deterministic) objects. The deterministic basis corresponds to a specification of our basic objects. Therefore, also the term rewriting calculus has to be adapted to respect the decision that variables range only over single values. So only deterministic terms can be substituted for a variable.

A first approach in the direction of a deterministic base could be to designate a subset of the operation symbols as the "basic operations". This idea is sufficient for many applications (and will be studied below in more detail), however it is a special case of a simpler approach. The idea is generally to fix a subset D of the terms which are "deterministic terms". If all terms in D are interpreted as singleton sets, we have a compatibility property with the inclusion rules:

$$\text{If } t \in D \text{ and } T \models t \to t', \text{ then } t' \in D .$$

The terms contained in D can be marked by writing down an axiom

$$DET(t) \qquad \text{(read: "t is deterministic").}$$

Then the compatibility property can be made into a deduction rule for such formulas. The set D then is described indirectly by

$$D = \{t \in W(\Sigma, X) \mid T \vdash DET(t)\}.$$

There is a close analogy between this idea and the extension of algebraic specifications to partial functions as it is explained in ([Broy,Wirsing82]) using a *definedness predicate*. Interested readers can find more detailed material on this topic in chapter 6.

2.1.2 Determinacy Predicate

The following definition just formalizes the concepts which were explained above informally.

Definition 2.2 (DET-Axiom, Validity)

A $(\Sigma, X\text{-})$ *DET-axiom* is a term, which is denoted as a formula using the so-called *determinacy predicate* or *DET-predicate* :

DET(t)

where $t \in W(\Sigma, X)$.

A DET-axiom ⟨DET(t)⟩ is *valid* in a Σ-Algebra A (A ⊨ DET(t)) iff for all valuations $\beta \in ENV(X, A)$ the interpretation is determinate:

$$| I_\beta^A[t] | = 1.$$

The notions "algebraic specification" and "model" from now on are meant to admit DET-axioms within the axiom set, too.   ◊

Example 2.3

Let the specification DOUBLE from example 1.14 be extended to a new specification DOUBLE' which contains the following additional axioms:

DET(zero),   DET(succ(x))

The algebra N from example 1.14 is a model of DOUBLE', too.

NN from example 1.15 is not a model of DOUBLE', since
$$| I^{NN}[zero] | = | \{ 0, 1 \} | = 2.$$
Moreover, in N the following formulae hold (which are not axioms):

N ⊨ DET(add(x,y))   N ⊨ DET(double(x))   ◊

The term rewriting calculus now is extended by deduction rules for DET-axioms. The deduction rule (AXIOM) is modified, in order to ensure soundness: Variables now can be instantiated only with such terms which are proven to be deterministic.

The calculus defined below is the most frequently used calculus in this text. Therefore deductions within this calculus are written without a special index, in difference to deductions within all other calculi (like $\vdash_{RC}$).

Definition 2.4 (Term Rewriting with DET)

Let $T = (\Sigma, R)$ be a nondeterministic algebraic specification (with DET-axioms). A formula $\langle t1 \to t2 \rangle$ or $\langle DET(t) \rangle$, respectively, is *deducible* in T, written symbolically:

$$T \vdash t1 \to t2 \qquad \text{or} \qquad T \vdash DET(t), \text{ respectively},$$

iff there is a deduction for the formula using the following deduction rules:

(REFL)
$$\frac{}{t \to t} \qquad \text{if } t \in W(\Sigma, X)$$

(TRANS)
$$\frac{t1 \to t2, \; t2 \to t3}{t1 \to t3} \qquad \text{if } t1, t2, t3 \in W(\Sigma, X)$$

(CONG)
$$\frac{t_i \to t_i{}'}{f(t_1, \ldots, t_{i-1}, t_i, t_{i+1}, \ldots, t_n) \; \to \; f(t_1, \ldots, t_{i-1}, t_i{}', t_{i+1}, \ldots, t_n)}$$

if $[f: s_1 \times \ldots \times s_n \to s] \in F$,
$t_j \in W(\Sigma, X)_{s_j}$ where $j \in \{1, \ldots, n\}, t_i{}' \in W(\Sigma, X)_{s_i}$

(AXIOM-1)
$$\frac{DET(\sigma x_1), \ldots, DET(\sigma x_n)}{\sigma l \to \sigma r}$$

if $\langle l \to r \rangle \in R, \sigma \in SUBST(\Sigma, X)$,
$\{x_1, \ldots, x_n\} = Vars(l) \cup Vars(r)$

(AXIOM-2)
$$\frac{DET(\sigma x_1), \ldots, DET(\sigma x_n)}{DET(\sigma t)}$$

if $\langle DET(t) \rangle \in R, \sigma \in SUBST(\Sigma, X)$,
$\{x_1, \ldots, x_n\} = Vars(t)$

(DET-X) $\dfrac{\quad\rule{2cm}{0.4pt}\quad}{DET(x)}$ if x ∈ X

(DET-D) $\dfrac{DET(t1),\ t1 \rightarrow t2}{DET(t2)}$ if t1, t2 ∈ W(Σ, X)

(DET-R) $\dfrac{DET(t1),\ t1 \rightarrow t2}{t2 \rightarrow t1}$ if t1, t2 ∈ W(Σ, X)   ◊

Example 2.5

Examples for deductions in the specification DOUBLE' of example 2.3:

DOUBLE' ⊢ double(zero) → zero :

(1) ⊢ DET(zero)	(AXIOM-2)
(2) ⊢ double(zero) → add(zero,zero)	(AXIOM-1), (1)
(3) ⊢ add(zero,zero) → zero	(AXIOM-1), (1)
(4) ⊢ double(zero) → zero	(TRANS), (2), (3)

DOUBLE' ⊢ double(zero_or_one) → succ(succ(zero)) :

(1) ⊢ DET(zero)   (AXIOM-2)
(2) ⊢ DET(succ(zero))   (AXIOM-2), (1)
(3) ⊢ add(succ(zero),succ(zero)) → succ(add(zero,succ(zero)))
                                                    (AXIOM-1), (1),(2)
(4) ⊢ add(zero,succ(zero)) → succ(zero)   (AXIOM-1), (2)
(5) ⊢ succ(add(zero,succ(zero))) → succ(succ(zero))
                                                    (CONG), (4)
(6) ⊢ add(succ(zero),succ(zero)) → succ(succ(zero)) (TRANS), (3), (5)
(7) ⊢ double(succ(zero)) → add(succ(zero),succ(zero))
                                                    (AXIOM-1), (2)
(8) ⊢ double(succ(zero)) → succ(succ(zero))   (TRANS), (7), (6)
(9) ⊢ zero_or_one → succ(zero)   (AXIOM-1)
(10) ⊢ double(zero_or_one) → double(succ(zero))   (CONG), (9)
(11) ⊢ double(zero_or_one) → succ(succ(zero))   (TRANS), (10), (8)

The "wrong" deduction from example 1.14 is not allowed here:
$$\neg (\, \text{DOUBLE' } \vdash \text{double(zero\_or\_one)} \rightarrow \text{succ(zero)} \,) \qquad \lozenge$$

Theorem 2.6 (Soundness)

Let $T = (\Sigma, R)$ be a nondeterministic algebraic specification on a deterministic basis. Then for $t, t1, t2 \in W(\Sigma, X)$:

$$T \vdash t1 \rightarrow t2 \quad \Rightarrow \quad Mod(T) \models t1 \rightarrow t2$$
$$T \vdash DET(t) \quad \Rightarrow \quad Mod(T) \models DET(t)$$

Proof: By induction on the derivation, see appendix A. $\qquad \lozenge$

The calculus above is slightly more restricted than it was necessary for soundness: The premises of deduction rule (AXIOM-1) are needed only for variables which have *multiple* occurrences in the right hand side of the axiom. This is a simple consequence of theorem 1.19. Therefore, the following deduction rule is sound, too:

(AXIOM-1-RLIN) $DET(\sigma x_1), ..., DET(\sigma x_n)$

$$\overline{\hspace{4cm}}$$

$$\sigma l \rightarrow \sigma r$$

if $\triangleleft l \rightarrow r \triangleright \in R, \sigma \in SUBST(\Sigma, X)$,
$\{ x_1, ..., x_n \} = \{ x \in Vars[r] \mid \exists u1, u2 \in Occ[r]: u1 \neq u2 \wedge r/u1 = x \wedge r/u2 = x \}$

The results given below can be obtained also using the calculus of definition 2.4, which is simpler in its structure. This is one reason why the calculus of definition 2.4 is preferred within this manuscript. The other reason is that there is no significant gain in completeness if the more complex rule (AXIOM-1-RLIN) is used. The next section studies completeness issues in more detail.

2.1.3 Completeness: A Negative Result

It is the obvious next question whether the calulus introduced above in Definition 2.4 is complete in some sense. Unfortunately, there exist counterexamples which demonstrate the incompleteness of the calculus.

For a restricted specification language (right-linear rules), in chapter 1 a completeness proof was given (theorem 1.23). However, the proof technique used there cannot be applied in the same way to specifications on a deterministic basis. If the deduction rule (AXIOM-1) is used to prove the validity of the axioms within a term model, then the carrier set of the term model contains provably deterministic terms. On the other hand, if we want to derive from $Mod(T) \models t1 \rightarrow t2$ a proof of the formula $t1 \rightarrow t2$ within the calculus, then the interpretation of the term $t1$ (within the term model) must contain the term $t2$ itself. This means, this idea leads at most to a proof for the following property:

(*) $\forall$ t1, t2∈W(Σ, X):

$$Mod(T) \models t1 \rightarrow t2 \ \wedge \ T \vdash DET(t2) \ \Rightarrow \ T \vdash t1 \rightarrow t2.$$

This property (*) is called *weak completeness* below.

Weak completeness still is an interesting result if looked at from the programmer's viewpoint. It states that for every pair of terms, denoting a nondeterministic expression and a value, the calculus provides a satisfactory method to check whether the value is a possible outcome of the nondeterministic expression within all models. Unfortunately, the following counterexample shows that even weak completeness does not hold in general.

Example 2.7

Consider the following nondeterministic specification INC:

spec INC
sort   s
func   a: $\rightarrow$ s, b: $\rightarrow$ s,
          f: s $\rightarrow$ s,   g: $\rightarrow$ s,
          h: s × s $\rightarrow$ s,   k: s $\rightarrow$ s
axioms
          DET(a),   DET(b),
          f(g) $\rightarrow$ a,   f(a) $\rightarrow$ b,   f(b) $\rightarrow$ b,
          g $\rightarrow$ b,   h(x,a) $\rightarrow$ a,   h(x,b) $\rightarrow$ b,
          k(x) $\rightarrow$ h(x,f(x))
end

We show now that

INC $\models$ k(g) $\rightarrow$ a,

i.e. that this inclusion holds in every model of the specification INC.

Let A be a model of INC. Then the interpretations of the terms a and b must be singleton sets; therefore we use the convention $I^A[a] = \{\ a\ \}$, $I^A[b] = \{\ b\ \}$.

The axiom

$$f(g) \rightarrow a$$

means that

$$I^A[f(g)] = \{\ e \in f^A(e') \mid e' \in g^A\ \} \supseteq \{\ a\ \}.$$

Therefore there is an element $e_0 \in g^A$ such that $a \in f^A(e_0)$. Using this element e_0, the

following chain of inclusions holds:

$$I^A[\ k(g)\ ]$$

$$\supseteq \quad k^A(e_0) \quad \text{(since } e_0 \in g^A, \text{ and because of Defn. 1.5)}$$

$$\supseteq \quad h^A(e_0, f^A(e_0)) \quad \text{(axiom } \langle k(x) \rightarrow h(x, f(x)) \rangle\text{)}$$

$$\supseteq \quad h^A(e_0, a) \quad \text{(because of } a \in f^A(e_0)\text{)}$$

$$\supseteq \quad \{\ a\ \} \quad \text{(axiom } \langle h(x, a) \rightarrow a \rangle\text{)}.$$

Using deductions within the calculus of Definition 2.4, this inclusion cannot be proven. The only way to reduce a term starting with a "k" is by applying the axiom $\langle k(x) \rightarrow h(x, f(x)) \rangle$. This axiom can be applied only, if a provably deterministic term is substituted for the variable x (for soundness reasons). The only provably deterministic terms are a and b, therefore we can deduce:

$$\text{INC} \vdash k(a) \rightarrow h(a, f(a)), \quad \text{INC} \vdash k(b) \rightarrow h(b, f(b)).$$

Only the second one of these inclusions can be connected with the term k(g) by the axiom $\langle g \rightarrow b \rangle$. Using (TRANS), we have:

$$\text{INC} \vdash k(g) \rightarrow h(b, f(b)).$$

Unfortunately, the only way to reduce the right hand side of this inclusion further is by deducing:

$$\text{INC} \vdash k(g) \rightarrow b \quad \text{(axiom } \langle h(x, b) \rightarrow b \rangle\text{)}.$$

There is no way to reach the term a by such a deduction:

$$\neg (\text{INC} \vdash k(g) \rightarrow a). \qquad\qquad \lozenge$$

The example 2.7 does not only show that the calculus is incomplete; it even shows that weak completeness does not hold. This follows simply from the fact that the inclusion used to demonstrate the incompleteness had a deterministic term on its right hand side. This example works as a counterexample also for another popular way of weakening the notion of completeness: the so-called *ground completeness*, where attention is restricted to inclusions between ground terms. (The terms k(g) and a used in the example are ground.)

Another interesting observation is that a relaxation of the calculus as indicated above, using the rule (AXIOM-1-RLIN), does not avoid the completeness problem. In this modified calculus it is also forbidden to instantiate the variable x in ⟨k(x) → h(x,f(x))⟩ to the term g.

At this point, again a decision must be made where to attack the deficiencies which were exposed by the example. There are two options: to change the calculus again or to restrict the syntactic form of the axioms (but not as severely as to right-linearity).

A closer inspection of the example above gives some hints how to decide. The problem in the example comes from the fact that the semantic argumentation mainly relies on the axiom

$$f(g) \to a,$$

which cannot be used in the deduction (since the term f(g) cannot be generated). Generally, this axiom has a somewhat spurious meaning. The following argumentation (which sloppily mixes syntax and semantics) tries to isolate the problem: The axiom says something about the meaning of f applied to g, but it does not explain the consequences for both single functions f and g. The only property of g directly stated in the axioms is

$$g \to b,$$

but this value of b for g apparently does not lead to a value of a for f(g) (since f(b) seems only to have the value b). Even if g had additionally the value of a, nothing would change here. So the axioms implicitly contain the assumption that there is another "third" basic object, let us call it c. This element c is distinct from a and b, c is a value of g and f applied to c delivers the value of a. This complex argumentation obviously does not fit into the simple framework of a rewriting-like calculus.

The framework of [Walicki 92/93] showed recently, how the semantic arguments can be transferred into a calculus. The main idea there is to introduce a "binding context" for variables. So the sentence "There is an element $e_0 \in g^A$ such that $a \in f^A(e_0)$" from the proof above is formalized as a deduction rule (binding introduction) which can derive the formula

$$x \in g \Rightarrow f(x) \to a.$$

The variable x in this formula still denotes a single value; the $\in$-sign and the implication are to be interpreted with their usual mathematical semantics. Based on this formula, the other axioms can be applied leading to

$$x \in g \Rightarrow k(x) \rightarrow a.$$

A final deduction (binding elimination) now can remove the binding, since the bound variable occurs only once within the term:

$$k(g) \rightarrow a.$$

In [Walicki 92/93], soundness and completeness of this more complex calculus is shown. This paper also contains a detailed comparison with our work. For the purposes of the text at hands, however, we will concentrate on rewriting-oriented and tool-supported calculi like the one from definition 2.4. In order to achieve completeness for this kind of calculi, we have to exclude the anomalies shown by example 2.7.

In the next sections we will restrict the axioms syntactically in such a way that the calculus directly can handle it. This does not necessarily mean to exclude axioms like ⟨f(g) → a⟩, where a nondeterministic function is applied to another one within the left hand side. But the restriction will ensure that such an axiom is consistent with some other deduction which shows how the computation can be led applying only functions to deterministic terms.

From a methodical point of view, it is important to state that specifications like INC above are not simply "bad". Such a specification must be seen as a rather abstract and sketchy formulation which just does not fix all details how the functions work together. The restrictions defined in the next section describe a smaller class of specifications which is suitable for a term rewriting style of deduction. This can be seen as a step from abstract specification towards programming.

2.2 Additive Specifications

This section shows how the specification language can be adapted more closely to the needs of a deterministic basis, such that a completeness result for the calculus from above holds.

It is a good starting point for this section to think about the way how a general term model for a specification with a deterministic basis can be constructed. From the idea underlying the notion of a deterministic basis, it is obvious that the carrier sets of such a term model must be formed by *provably deterministic terms*. The natural interpretation for a given nondeterministic term then is the set of all deterministic terms it can be reduced to. In order to ensure the well-definedness of such a model, two properties must be fulfilled:

- For every nondeterministic term there must be a deterministic term it can be reduced to. This ensures that the interpretation of every term is a nonempty set. This property is called *DET-completeness* below.

- The effect which was present in the example 2.7 from above must be avoided. This property is called *DET-additivity* below.

2.2.1 DET-Completeness and DET-Additivity

The first and rather simple condition for the construction of a term model is DET-completeness. Formally, it means:

$$\forall t: \exists t': \ T \vdash t \rightarrow t' \ \wedge \ T \vdash DET(t').$$

DET-completeness is very similar to the so-called *sufficient completeness* known from the classical theory of algebraic specifications [Guttag 75]. This similarity helps to make precise the ranges of the quantifiers which have been omitted in the formula above. It is reasonable to restrict the range for t and t' to *ground terms*. Otherwise, for every term containing variables (like add(x,y)) there must be a deterministic term it can be reduced to! This would definitely be a too strong restriction for practical specifications. At this point, it becomes obvious that the term model will be constructed also from *ground deterministic* terms only, and therefore will only help to ensure ground completeness.

Please note that due to the similarity of the notions, the existing methods for testing sufficient completeness can be carried over for testing DET-completeness, too (see also sections 2.4 and 4.4.1).

Definition 2.8 (DET-Completeness)

A specification T = (Σ, R) over a deterministic basis is called *DET-complete* iff

$$\forall t \in W(\Sigma): \exists t' \in W(\Sigma): \ T \vdash t \rightarrow t' \ \wedge \ T \vdash DET(t'). \qquad \Diamond$$

The second, more complex notion to be defined is DET-additivity. It is understood best by looking again at the problematic axiom from example 2.7:

$$f(g) \rightarrow a.$$

This is an inclusion which is deducible (since it is an axiom), but which is not consistent with the inclusions holding in a term model. Within a term model D constructed from deterministic terms, the interpretation of $f(g)$ is defined additively:

$$I^D[\, f(g) \,] = \{\, t \in f^D(t') \mid t' \in I^D[\, g \,] \,\}.$$

Using the axioms of specification INC, and the convention that a term is interpreted by the deterministic terms it can be reduced to, the interpretation of g must be:

$$I^D[\, g \,] = \{\, b \,\}.$$

Again, using the axioms of INC, it is not possible to reduce the term $f(b)$ to a. So the deterministic term a will not be contained within $f^D(b)$, and also not in $I^D[\, f(g) \,]$.

This means that the axiom $\langle f(g) \rightarrow a \rangle$ states a *non-additive* property, which cannot be derived by first looking at the interpretation of the arguments and then at the operation applied to them. The property which is necessary for an additive axiom system is, for this example:

$$\exists\, t: \vdash DET(t) \wedge \vdash g \rightarrow t \wedge \vdash f(t) \rightarrow a.$$

Obviously, INC does not fulfil this property. A generalization to terms with arbitrary many arguments gives the formal definition of DET-additivity.

Definition 2.9 (DET-Additivity)

A specification $T = (\Sigma, R)$ over a deterministic basis is called *DET-additive* iff

$$\forall [f: s_1 \times \ldots \times s_n \rightarrow s] \in F:$$
$$\forall\, t_1 \in W(\Sigma)_{s_1}, \ldots, t_n \in W(\Sigma)_{s_n}, t \in W(\Sigma)_s:$$
$$T \vdash f(t_1, \ldots, t_n) \rightarrow t \wedge T \vdash DET(t) \Rightarrow$$
$$\exists\, t_1' \in W(\Sigma)_{s_1}, \ldots, t_n' \in W(\Sigma)_{s_n}:$$
$$T \vdash f(t_1', \ldots, t_n') \rightarrow t \wedge$$
$$T \vdash t_1 \rightarrow t_1' \wedge \ldots \wedge T \vdash t_n \rightarrow t_n' \wedge$$
$$T \vdash DET(t_1') \wedge \ldots \wedge T \vdash DET(t_n') \qquad \Diamond$$

DET-additivity means that the term rewriting relation is an additive extension of rewriting on deterministic terms. In other words, the specification must be equivalent to a set R ' of inclusion axioms

$f(t_1',...,t_n') \rightarrow t'$

where $t_1',..., t_n', t'$ are deterministic terms.

Example 2.10

The specification INC from example 2.7 is DET-complete, but not DET-additive. It can be made DET-additive by adding the axiom

$f(b) \rightarrow a$.

There are many other ways to achieve DET-completeness, among them an extension of the signature by a new constant $c: \rightarrow s$, with the new axioms:

DET(c), $g \rightarrow c$, $f(c) \rightarrow a$.   ◊

In a more abstract view, the DET-additivity of a specification means that non-determinism is specified in a *local* manner, that is as a number of alternatives for the behaviour of a single function. The specification INC, however, contains a kind of "global" nondeterminism which does not belong to either the function f or g (but to the collaboration of both). This sort of effect is called "non context-free nondeterminism" in [Kaplan 88]. Similar to our approach, [Kaplan 88] excludes the unwanted form of nondeterminism by a restriction to so-called regular specifications. The main advantage of DET-additivity, as it is defined here, over regularity is that DET-additivity immediately ensures a kind of completeness for the rewriting calculus. In regular specifications, a particular kind of confluence is needed again for completeness of term rewriting. For the DET-additivity of a specification there is a rather simple criterion which can be used in many practical examples:

Theorem 2.11

If a specification $T = (\Sigma, R)$ fulfils the conditions A1 and A2 below, then T is DET-additive:

(A1)   For all axioms $l \rightarrow r \in R$, the term l does not consist of a single variable, i.e. $l = f(t_1,...,t_n)$.
          Moreover, for all $i \in \{1,..., n\}$: $T \vdash DET(t_i)$.

(A2)   For all deterministic terms t (i.e. T |- DET(t)), where t does
        not consist of a single variable (i.e. $t = f(t_1,...,t_n)$), the
        subterms must be deterministic again, i.e. for all $i \in \{1, ..., n\}$:
        T |- DET(t_i).

Proof: See appendix A.   ◊

The specification DOUBLE' from example 2.3 can be proven to be DET-
additive using theorem 2.11. Chapters 5 and 7 contain larger examples which
show the practical application of the criterion.

2.2.2 Term Models and Completeness

Now the construction of the term model can be given in detail, which was the
main motivation for introducing the notions of DET-completeness and DET-
additivity. The following notion is a preliminary for the model construction:

Definition 2.12 (Induced Equivalence of Terms)

A nondeterministic algebraic specification $T = (\Sigma, R)$ induces a relation
$\approx$ on $W(\Sigma)$ as follows
$$t1 \approx t2 \quad \Leftrightarrow_{def} \quad T \mid\text{-} t1 \rightarrow t2 \ \wedge \ T \mid\text{-} t2 \rightarrow t1$$
(where $t1, t2 \in W(\Sigma)$).

The deduction rules of the calculus (Definition 2.4) ensure that $\approx$ is an
equivalence relation as well as a congruence with respect to the term-
constructing operations. [t] denotes the equivalence class of the term t
with respect to $\approx$.   ◊

The construction of a term model now uses equivalence classes with respect to $\approx$
as its carriers.

Definition 2.13 (Term Model DΣ/R)

Let $T = (\Sigma, R)$ be a DET-complete specification. The algebra DΣ/R is
defined by:
$$_s D\Sigma/R = \{[t] \mid t \in W(\Sigma) \ \wedge \ T \mid\text{-} DET(t) \} \quad \text{where } s \in S$$

$$f^{D\Sigma/R} \colon W(\Sigma)_{s_1}/\approx \times \ldots \times W(\Sigma)_{s_n}/\approx \to \wp^+(W(\Sigma)_s/\approx)$$

$$f^{D\Sigma/R}([t_1],\ldots,[t_n]) =$$

$$\{[t] \mid t \in W(\Sigma) \wedge T \vdash DET(t) \wedge T \vdash f(t_1,\ldots,t_n) \to t \}$$

where $[f \colon s_1 \times \ldots \times s_n \to s] \in F$.

The DET-completeness ensures that $f^{D\Sigma/R}([t_1],\ldots,[t_n]) \neq \emptyset$ and $s^{D\Sigma/R} \neq \emptyset$ (since Σ is presupposed to be sensible). The well-definedness of DΣ/R follows from the fact that $\approx$ is a congruence.   ◊

This algebra corresponds well to the intuitive understanding of a nondeterministic specification. The algebra DΣ/DOUBLE' according to example 2.3, for instance, is isomorphic to the model N from example 1.14. However, the algebra DΣ/INC according to example 2.7 is *not* a model of INC. In order to ensure that the term algebra really belongs to the model class, the property of DET-additivity is needed.

Theorem 2.14

Given a DET-complete und DET-additive specification $T = (\Sigma, R)$, the algebra DΣ/R according to definition 2.13 is a model of T.

Proof: See appendix A.   ◊

The main reason for the construction of the term model was to prove a completeness result. This result is formulated within the corollary below. The term model is needed also for another sort of results, which refer to initiality. For such results, see chapter 3.

The kind of completeness which follows from the term model construction is restricted in two ways:

- It refers only to inclusions between ground terms, since the term model uses ground terms for its carrier sets. (This is due to the fact that we did not want to impose a too strong version of the DET-completeness property on the specifications.)
- It refers only to inclusions which have a deterministic term as their right hand side. This was called weak completeness above, and is a consequence of the fact that the model uses only deterministic terms for its carrier sets. (This is the price which has to be paid for the soundness of the calculus.)

Corollary 2.15   **(Weak Ground Completeness)**

Let T = (Σ, R) be a DET-complete and DET-additive specification, A∈Mod(T). Then for t1, t2∈W(Σ):

$$\text{Mod(T)} \models t1 \rightarrow t2 \ \wedge \ T \vdash \text{DET(t2)} \ \Rightarrow \ T \vdash t1 \rightarrow t2$$

Proof:

Mod(T) ⊨ t1 → t2

⇒   DΣ/R ⊨ t1 → t2   (Theorem 2.14)

⇒   (∀ t': T ⊢ DET(t') ∧ T ⊢ t2 → t' ⇒ T ⊢ t1 → t')
                                (Lemma 2.14.1, see appendix A)

⇒   T ⊢ t1 → t2   (because of T ⊢ DET(t2), using (REFL))   ◊

The following counterexample illustrates the fact that only ground weak completeness has been achieved. It shows that in general ground completeness does not hold even for DET-complete and DET-additive specifications.

Example 2.16

> **spec** GIC
> **sort** s
> **func**   a: → s,   b: → s,
>   g: → s,   f: s → s,   h: s × s → s
> **axioms**
>   DET(a),   DET(b),
>   g → a,   f(x) → h(x,x),   h(x,x) → x
> **end**

For an arbitrary A∈Mod(IC), semantic arguments show that A ⊨ f(g) → g :

$e \in g^A$

$\Rightarrow e \in I_\beta^A[x]$ where β(x) = e   (Definition 1.5)

$\Rightarrow e \in I_\beta^A[h(x,x)]$ (because of h(x,x) → x)

$\Rightarrow e \in I_\beta^A[f(x)]$ (because of f(x) → h(x,x))

$\Rightarrow e \in \{k \in f^A(l) \mid l \in I_\beta^A[x]\}$ (Definition 1.5)

$$\Rightarrow e \in f^A(e) \qquad\qquad\qquad (I_\beta^A[x] = \{e\})$$

$$\Rightarrow e \in \{k \in f^A(l) \mid l \in I_\beta^A[g]\} \qquad (\text{because of } e \in g^A)$$

$$\Rightarrow e \in I_\beta^A[f(g)] \qquad\qquad\quad (\text{Definition 1.5}).$$

But this inclusion is *not* deducible, since DET(g) is not deducible in IC. Even if the calculus is extended by the deduction rule (AXIOM-1-RLIN), the inclusion $f(g) \to g$ cannot be deduced from the axioms of GIC.   ◊

The next section aims at a situation where a true (non-weak) ground completeness result can be shown. This leads to a final refinement of the concepts, concerning the calculus as well as the model classes.

2.3 Junk-Free Models

This section concludes the investigation of completeness results by showing how the restriction to "weak" completeness can be removed. It is shown that this can be achieved by similar techniques as they are used for the treatment of term-generated models in the classical case.

2.3.1 "Junk" in Nondeterministic Models

The notion of "junk" is well known from the theory of equational specifications. There it is used to denote elelents within the carrier set of a model which are not an image of the interpretation of some term. Such elements cannot be constructed by the provided operations, and they cannot be controlled by deductions using terms over the given signature. However, propositions containing free variables always have a semantics where the variables also range over junk elements. It is widely accepted that a practically usable specification language has to concentrate on models which do not contain junk. In particular, for junk-free models it is sound to use an induction principle on the structure of terms, which is one of the most important proof techniques in the field of program and data structure verification. The semantical investigations in this manuscript also aim at junk-free models.

It is an interesting observation that nondeterminism introduces a second source of junk besides the classical problem concerning the range of free variables. In nondeterministic specifications, there exists also a dimension which is called the *breadth* of a nondeterministic expression. The breadth is the range of possible outcomes for a nondeterministic computation. Some observations clearly indicate a similarity between "non-standard elements" in the classical junk priciple and "non-standard outcomes of a nondeterministic expression". As an illustration, the example 2.16 is revisited.

Example 2.17

In example 2.16, the following specification has been defined:

spec GIC
sort s
func   a: → s,   b: → s,
           g: → s,   f: s → s,   h: s × s → s
axioms
           DET(a),   DET(b),
           g → a,   f(x) → h(x,x),   h(x,x) → x
end

The term model $D\Sigma$/GIC uses the following interpretation:
$$I^{D\Sigma/R}[\,g\,] = \{\,[\,a\,]\,\}, \quad I^{D\Sigma/R}[\,a\,] = \{\,[\,a\,]\,\},$$
$$I^{D\Sigma/R}[\,f(g)\,] = \{\,f^{D\Sigma/R}(a)\,\} = \{\,[\,a\,]\,\}.$$
Therefore in $D\Sigma$/GIC the following inclusions are valid:
$$D\Sigma/R \models f(g) \to g, \qquad D\Sigma/R \models a \to g.$$

A different model M of GIC is given by
$$s^M = \{\,a, b\,\}, \quad a^M = \{\,a\,\}, \quad b^M = \{\,b\,\}, \quad g^M = \{\,a\,,b\,\},$$
$$f^M(e) = \{\,e\,\}, \quad h^M(e1,e2) = \{\,e1\,\} \quad \text{for } e, e1, e2 \in \{a, b\,\}.$$

The model M uses the following interpretation:
$$I^M[\,g\,] = \{\,a, b\,\}, \qquad I^{D\Sigma/R}[\,a\,] = \{\,a\,\},$$
$$I^M[\,f(g)\,] = \{\,a, b\,\}.$$

Therefore in M the inclusion $\langle f(g) \rightarrow g \rangle$ is valid ($M \models f(g) \rightarrow g$), but $\langle a \rightarrow g \rangle$ is not valid ($\neg$ ($M \models a \rightarrow g$)).   ◊

From this example, some observations can be made:

(a)   Within the model $D\Sigma/R$, a number of inclusions (even ground inclusions) hold, which cannot be deduced by the calculus of Definition 2.4. As an example, consider the inclusion $\langle f(g) \rightarrow g \rangle$ which does hold in $D\Sigma/GIC$, but which is not deducible.

(b)   Within the model $D\Sigma/R$, a number of inclusions (even ground inclusions) hold, which do not hold in all models. As an example, consider the inclusion $\langle a \rightarrow g \rangle$ which does hold in $D\Sigma/GIC$, but not in M.

(c)   The phenomena described above appear only for inclusions the right hand side of which is not provably deterministic. (For other inclusions, Corollary 2.15 can be applied.)

This situation is quite similar to the situation in classical equational logic where a ground term model can be constructed also (the so-called initial model). The analogy is obvious:

(a')   Within the initial model, some equations hold, which cannot be deduced by equational reasoning. These equations are called "inductive consequences".

(b')   Within the initial model, some equations hold, which do not hold in all models. This also refers to the "inductive consequences", which do hold only for the so-called term-generated (junk-free) models.

(c')   The phenomena described above appear only for equations which contain free variables (non-ground equations).

Within nondeterministic specifications, both difficulties arise. The sort of difficulty described by (a) to (c) exists *even for ground inclusions*. This is due to unexpected elements in the breadth of a nondeterministic expression (like b in $I^M[g]$). Obviously, the difficulty concerning non-ground inclusions (as in (a') to (c')) is present within nondeterministic specifications, independently of that.

In the following, the techniques known from the classical theory for the treatment of junk are carried over to treat the problem of junk in the nondeterministic breadth of a term (see (a) to (c) from above). For this purpose,

(1) The calculus is extended to a kind of "inductive" calculus which describes exactly those inclusions which are valid in the term model $D\Sigma/R$;

(2) The model class is restricted to junk-free models in such a way that the extended calculus is sound and model M from the example above is excluded.

2.3.2 Breadth Induction

Within the theory of equational specifications, there is a calculus for the deduction of so-called "inductive" consequences which are valid only for junk-free models. The basic idea of the calculus is to describe exactly the equalities within the initial model. Due to Gödel's results, such a calculus is either incomplete or it is different from a true formal system (since the theory of Peano arithmetics can be described by initial models). A rather well-known technique for such a calculus is the use of a *semi-formal system*. This means that deduction rules are used which have an infinite number of premises. For practical proofs, the infinite premise is covered by a kind of induction proofs (for instance on the term structure). Below an extension of the calculus from definition 2.4 is given which also contains semi-formal rules.

Definition 2.18 (Breadth Induction Calculus)

The calculus given by definition 2.4 is extended by the following semi-formal rules:

(IND-R)
$$\frac{\forall\ t{\in}W(\Sigma):\ \vdash DET(t)\ \wedge\ \vdash t2 \to t\ \Rightarrow\ \vdash t1 \to t}{\vdash_{IND}\ t1 \to t2}$$

if $t1, t2 \in W(\Sigma)$

(IND-D)

$$\forall\ t1, t2 \in W(\Sigma):\ \vdash t \to t1\ \wedge\ \vdash DET(t1)\ \wedge\ \vdash t \to t2\ \wedge\ \vdash DET(t2)$$
$$\Rightarrow\ \vdash t1 \to t2$$

$$\vdash_{IND} DET(t)$$

if $t \in W(\Sigma)$

The calculus is called "inductive", and its derivations are denoted using the symbol $\vdash_{IND}$, since in many cases the premises of the rules can be proven only using an induction principle. This is the case as soon as the number of deterministic terms a given term can be reduced to is infinite (so-called *unbounded nondeterminism*). Please note that for specifications containing only bounded nondeterminism, the "inductive" calculus remains a formal system.

The following theorem shows that breadth induction for ground inclusions exactly deduces the inclusions valid in the term model $D\Sigma/R$.

Theorem 2.19 (Correspondence with the Term Model)

Let $T = (\Sigma, R)$ be a DET-complete and DET-additive specification. Then for ground terms $t, t1, t2 \in W(\Sigma)$:

$$T \vdash_{IND} t1 \to t2 \qquad \Leftrightarrow \qquad D\Sigma/R \models t1 \to t2$$
$$T \vdash_{IND} DET(t) \quad \Leftrightarrow \quad D\Sigma/R \models DET(t)\ .$$

Proof:

$D\Sigma/R \models t1 \to t2$

$\Leftrightarrow I^{D\Sigma/R}_{[t1]} \supseteq I^{D\Sigma/R}_{[t2]}$ (since $D\Sigma/R$ model of T)

$\Leftrightarrow \forall\ t \in W(\Sigma):\ \vdash DET(t)\ \wedge\ \vdash t2 \to t\ \Rightarrow\ \vdash t1 \to t$ (lemma 2.14.1)

$\Leftrightarrow \vdash_{IND} t1 \to t2$ (rule (IND-R))

$D\Sigma/R \models DET(t)$

$\Leftrightarrow |\ I^{D\Sigma/R}_{[t]}\ | = 1$ (since $D\Sigma/R$ model of T)

$\Leftrightarrow |\ \{\ [t']\ |\ \vdash t \to t'\ \wedge\ \vdash DET(t')\ \}\ | = 1$ (lemma 2.14.1)

$\Leftrightarrow \forall\ t1, t2 \in W(\Sigma):$

       $\vdash t \to t1\ \wedge\ \vdash DET(t1)\ \wedge\ \vdash t \to t2\ \wedge\ \vdash DET(t2) \Rightarrow \vdash t1 \to t2$

$\Leftrightarrow \vdash_{IND} DET(t)$ (rule (IND-D))   ◊

The example below shows a case where induction is used for proving the premise of a semi-formal rule:

Example 2.20

> **spec** INAT
> **sort**   Nat
> **func**   zero: $\to$ Nat,   succ: Nat $\to$ Nat,
>   double: Nat $\to$ Nat,   some: $\to$ Nat
> **axioms**
>   DET(zero),   DET(succ(x)),
>   double(zero) $\to$ zero,
>   double(succ(x)) $\to$ succ(succ(double(x)))
>   some $\to$ zero,
>   some $\to$ succ(some)
> **end**

We want to prove: MGen(T) $\models$ some $\to$ double(some). For this purpose, it can be proven (using structural induction) that for an arbitrary ground term $t \in W(\Sigma)$ fulfilling $\vdash$ DET(t) :

> $\vdash$ some $\to$ t

<u>t = zero:</u>
(1) $\vdash$ some $\to$ zero   (AXIOM-1)

<u>t = succ(t1):</u>
(1) $\vdash$ some $\to$ t1   (Induction hypothesis)
(2) $\vdash$ succ(some) $\to$ succ(t1)   (CONG)
(3) $\vdash$ some $\to$ succ(some)   (AXIOM-1)
(4) $\vdash$ some $\to$ succ(t1)   (TRANS), (3), (2)

Therefore:
$\forall$ t$\in$W(Σ): $\vdash$ double(some) $\to$ t $\wedge$ $\vdash$ DET(t) $\Rightarrow$ $\vdash$ some $\to$ t
and, using (IND-R):

> $\vdash_{IND}$ some $\to$ double(some). $\lozenge$

The calculus achieved so far is sound only for the standard term model, but it is not sound for arbitrary models, as can be seen from example 2.17. Breadth induction allows us to deduce the formula GIC $\vdash_{IND}$ DET(g), which does not hold in the model M. The next section excludes M as a model containing (breadth-)junk.

2.3.4 DET-Generated Models

In this section, a characterization for a class of models is given which obey a no-junk principle for the breadth of a nondeterministic term. This characterization mainly says that every possible deterministic outcome of a nondeterministic term must be due to a deduction within the specification. This leads to a formulation which may look a bit strange from the logical point of view, because it somehow mixes semantic and syntactic arguments. In section 3, a purely semantical characterization of junk-free models for nondeterministic specifications will be given.

Definition 2.21 (Term-Generation, DET-Generation)

A Σ-multi-algebra A is called *term-generated*, iff for all s $\in$ S:
$$\forall \, e \in s^A \colon \exists \, t \in W(\Sigma)_s \colon I^A[t] = \{ \, e \, \}$$

Gen(Σ) denotes the class of the term-generated Σ-multi-algebras. The class of all term-generated models of a specification T is called Gen(T).

Let T = (Σ, R) be a nondeterministic specification. A model A $\in$ Mod(T) is called *DET-generated*, iff for all s $\in$ S:
$$\forall \, t \in W(\Sigma)_s \colon \, \forall \, e \in I^A[t] \colon$$
$$\exists \, t' \in W(\Sigma)_s \colon \, T \vdash t \to t' \, \wedge \, T \vdash DET(t') \, \wedge \, I^A[t'] = \{ \, e \, \}.$$
The class of all DET-generated models of T is called DMod(T).   ◊

Example 2.22

By definition, the model DΣ/R for any DET-complete and DET-additive specification is DET-generated.

The model M from example 2.17 is not DET-generated, since b$\in$I^M[g], but this cannot be motivated by a deduction: $\neg$ (GIC $\vdash$ g $\to$ b).   ◊

The following theorem shows that the DET-generated models are exactly those models for which the extended calculus is sound and correct.

Theorem 2.23

Let $T = (\Sigma, R)$ be a DET-complete and DET-additive specification.
Then for t, t1, t2 $\in$ W(Σ):

$$D\Sigma/R \models t1 \rightarrow t2 \qquad \Leftrightarrow \qquad DMod(T) \models t1 \rightarrow t2,$$
$$D\Sigma/R \models DET(t) \qquad \Leftrightarrow \qquad DMod(T) \models DET(t).$$

Proof:

The "$\Leftarrow$"-direction follows from $D\Sigma/R \in DModT)$.

The first line of the "$\Rightarrow$"-direction can be seen as follows. Let A $\in$ DMod(T).

$e \in I^A[t2]$

$\Rightarrow \exists t': \vdash t2 \rightarrow t' \wedge \vdash DET(t') \wedge I^A[t'] = \{ e \}$ (A $\in$ DMod(T))

$\Rightarrow t' \in I^{D\Sigma/R}[t2]$ (Lemma 2.14.1)

$\Rightarrow t' \in I^{D\Sigma/R}[t1]$ ($D\Sigma/R \in Mod(T)$)

$\Rightarrow \vdash t1 \rightarrow t' \wedge \vdash DET(t') \wedge I^A[t'] = \{ e \}$ (Lemma 2.14.1)

$\Rightarrow e \in I^A[t1]$ (Theorem 2.6)

Therefore, A $\models$ t1 $\rightarrow$ t2. Analoguous arguments apply for DET(t). $\lozenge$

As already indicated, ground soundness and completeness now follows as a simple combination of the two last theorems.

Corollary 2.24 (Ground Soundness and Completeness)

Let $T = (\Sigma, R)$ be a DET-complete and DET-additive specification.

Then for ground terms t, t1, t2 $\in$ W(Σ, X):

$$T \vdash_{IND} t1 \rightarrow t2 \qquad \Leftrightarrow \qquad DMod(T) \models t1 \rightarrow t2,$$
$$T \vdash_{IND} DET(t) \qquad \Leftrightarrow \qquad DMod(T) \models DET(t).$$

Proof: Combination of theorems 2.19 and 2.23. $\lozenge$

2.3.5 Term-Generated Models

Before proceeding further, let us summarize what has been achieved so far:

- Weak Ground Soundness and Completeness (Corollary 2.15) for the calculus from defintion 2.4 and the class of all models;
- General Ground Soundness and Completeness (Corollary 2.24) for the calculus from definition 2.18 and the class of DET-generated models.

These results deal with the first source of junk (breadth-junk) in nondeterministic algebras. However, the standard model $D\Sigma/R$ is term-generated, too, and so is junk-free also with respect to the scond source. As an illustration, consider the following example:

Example 2.25

> **spec** NTG
> **sort** s
> **func** a: $\to$ s, b: $\to$ s, f: s $\to$ s
> **axioms**
> > DET(a), DET(b),
> > f(a) $\to$ a, f(b) $\to$ a
> **end**

> A model J for this specification is given by:
> > $s^J = \{ a, b, c \}$,
> > $f^J(a) = \{ a \}, f^J(b) = \{ a \}, f^J(c) = \{ c \}$.
> This model J is DET-generated but not term-generated.
> The inclusion
> > $f(x) \to a$
> does hold in $D\Sigma/NTG$ and in all term-generated models, but not within the model J (which contains the junk element c).   ◊

This example demonstrates clearly what the model class is which coincides best with the standard model $D\Sigma/R$. It is the class of *term-generated and DET-generated models*.

Definition 2.26

> Let T = (Σ, R) be a nondeterministic specification. The class of all DET-generated and term-generated models of T is called DGen(T).   ◊

One important observation is that the ground completeness results from above easily are carried over to the model class DGen(T).

Corollary 2.27 (Ground Completeness)

Let $T = (\Sigma, R)$ be a DET-complete and DET-additive specification.

Then for ground terms $t, t1, t2 \in W(\Sigma, X)$:

$$DGen(T) \models t1 \to t2 \quad \Rightarrow \quad T \vdash_{IND} t1 \to t2,$$
$$DGen(T) \models DET(t) \quad \Rightarrow \quad T \vdash_{IND} DET(t) .$$

Proof: Consequence of the fact that $D\Sigma/R \in DGen(T)$ and theorem 2.19. $\Diamond$

Another slightly more general result shows that within the model $D\Sigma/R$ exactly those *non-ground* inclusions hold which hold in the class $DGen(T)$. But please note that we did not give a calculus for generally deducing the non-ground inclusions which hold in $D\Sigma/R$. Such deductions may involve structural induction.

Theorem 2.28

Let $T = (\Sigma, R)$ be a DET-complete and DET-additive specification.

Then for $t, t1, t2 \in W(\Sigma, X)$:

$$D\Sigma/R \models t1 \to t2 \quad \Leftrightarrow \quad DGen(T) \models t1 \to t2,$$
$$D\Sigma/R \models DET(t) \quad \Leftrightarrow \quad DGen(T) \models DET(t).$$

Proof:

The "$\Leftarrow$"-direction follows from $D\Sigma/R \in DGen(T)$.

For the "$\Rightarrow$"-direction, let β be a valuation in $A \in DGen(T)$. Since A is term-generated and T is DET-complete, there is a substitution $\sigma \in SUBST(\Sigma)$ where: $\beta(x) = I^A[\sigma x]$ and $\vdash DET(\sigma x)$ for $x \in X$. Then $I^A_\beta[t] = I^A[\sigma t]$ for $t \in W(\Sigma, X)$, hence:

$$e \in I^A_\beta[t2] \Rightarrow e \in I^A[\sigma t2] \Rightarrow e \in I^A[\sigma t1] \quad \text{(as in theorem 2.23)}$$

$$\Rightarrow e \in I^A_\beta[t1].$$

Therefore, $A \models t1 \to t2$. Analoguous arguments apply for $DET(t)$. $\Diamond$

In some sense, with the class DGen(T) now a satisfactory semantics for nondeterministic specifications has been reached. The models of this class coincide in their important properties with the "standard" model dΣ/R. We have a simple weakly ground-complete calculus as well as a more complex ground-complete calculus available.

Therefore we turn now to the question, how the developed framework can be compared and integrated with other existing formalisms. This leads to several blocks of material which may be of varying interest for various readers. Only for readers which are interested in semantic considerations and generalizations of the field of algebraic specifications it is recommended just to follow the thread of the text. For readers interested in deduction- and programming-oriented aspects it may be a good idea to move on directly to chapter 4.

The following short section 2.4 gives a sketch how modularization techniques, as they have been developed in the field of algebraic specificaions, can be integrated with the nondeterministic framework.

Chapter 3 then presents a number of results on the model-theoretic semantics of nondeterministic specifications. These results are connected with the material from above mainly by two aspects:

- It is shown that the term model DΣ/R is an initial model within the class DGen(T). This can be seen as an additional argument showing that the "right" design decisions have been made.

- A more semantical characterization for the class DGen(T) is given, which avoids the "mixture" between syntactic and semantic notions used in definition 2.21.

Chapter 4 covers more general aspects by comparing nondeterministic specifications with common concepts like equational logic and logic programming.

2.4 Hierarchical Specifications

If a specification language is applied practically for the description of a larger system, means for *structuring* the whole text become very important. It is an advantage of algebraic specifications that there are criteria available which distinguish "good" modularizations. A "good" modular structure means here a structure where parts can be easily exchanged or refined without affecting other parts of the system. [Wirsing et al. 83] gives a detailed study of so-called hierarchical algebraic specifications. Below a short sketch is given, how the most important definitions and results concerning hierarchies can be transferred to the nondeterministic case.

Definition 2.29 (Hierarchical Specification)

A nondeterministic algebraic specification $T = (\Sigma, R)$ is called *hierarchical*, iff a subspecification T0 of T (i.e. $T0 = (\Sigma0, R0)$, $\Sigma0 \subseteq \Sigma$, $R0 \subseteq R$) is designated, which is called the *primitive* part of T.
A model $A \in DGen(T)$ is called hierarchical, iff the $\Sigma0$-reduct of A is in DGen(T0).

The specification T is called
- *hierarchy-preserving*, iff every model of A is hierarchical,
- *hierarchy-faithful*, iff every model $A0 \in DGen(T0)$ can be extended to a model $A \in DGen(T)$ such that the $\Sigma0$-reduct of A is A0.
- *hierarchy-persistent*, iff T is both hierarchy-preserving and hierarchy-faithful. $\Diamond$

As a syntactical representation of hierarchical specifications, we use a notation which is similar to [CIP85]. If the body of a specification contains a statement of the form
    basedon $P_1, ..., P_n$,
the union of $P_1, ..., P_n$ is meant to be the primitive part T0 of T.

The hierarchy-persistency of a specification in practice means that the primitive part and the non-primitive part can be developed independently, therefore it constitutes an important *modularity condition*. However, in order to check these conditions, we need more syntactical formulations. The following definition transfers the modularity conditions to the level of deductions.

Definition 2.30 (Hierarchy Conditions)

A hierarchical specification $T = (\Sigma, R)$ over a deterministic basis containing the primitive part $T0 = (\Sigma0, R0)$, $\Sigma0 = (S0, F0)$ is called *sufficiently complete*, iff:

$\forall\, t \in W(\Sigma)_s$:

$$T \vdash DET(t) \wedge s \in S0 \;\Rightarrow\; \exists\, t' \in W(\Sigma0): T \vdash t \rightarrow t' \,.$$

T is called *hierarchy-consistent* (sometimes also called *hiererchy-conservative*), iff:

$\forall\, t, t' \in W(\Sigma0)$:

$$T \vdash t \rightarrow t' \wedge T \vdash DET(t') \;\Rightarrow\; T0 \vdash t \rightarrow t' \wedge T0 \vdash DET(t').\Diamond$$

The following theorem shows (in analogy to a similar result in [Wirsing et al. 83] that these deductive properties ensure the semantic condition of hierarchy-persistency.

Theorem 2.31

Let $T = (\Sigma, R)$ be a hierarchical specification over a deterministic basis containing the primitive part $T0 = (\Sigma0, R0)$, $\Sigma0 = (S0, F0)$.

If T is sufficiently complete and hierarchy-consistent, then T is hierarchy-preserving.

Proof:

As a first step, we show that T is hierarchy-preserving. Let $A \in DGen(T)$. It is obvious that the $\Sigma0$-reduct of A also fulfils the axioms, so we have to show that the reduct is term- and DET-generated. For any element $e \in s0^A$ of a primitive carrier set ($s0 \in S0$), there is a term $t \in W(\Sigma)$ such that $I^A[t] = \{\, e\, \}$ (term generation of A). Because of the DET-generation of A, there is also a $t' \in W(\Sigma)$ such that $T \vdash DET(t')$, $T \vdash t \rightarrow t'$, and $I^A[t'] = \{\, e\, \}$. The sufficient completeness of T gives a term $t'' \in W(\Sigma0)$ such that $T \vdash t' \rightarrow t''$; obviously also $T \vdash DET(t'')$ and $I^A[t''] = \{\, e\, \}$. So term-generation of the reduct holds. Using (TRANS), we also have $T \vdash t \rightarrow t''$ and $T \vdash DET(t'')$, so by hierarchy-consistency of T also $T0 \vdash t \rightarrow t''$ and $T0 \vdash DET(t'')$, i.e. DET-generation holds, too. $\Diamond$

It is an interesting observation that this result needs a slightly stronger precondition than the corresponding proposition 4 of [Wirsing et al. 83].

Sufficient completeness alone ensures only the term-generation of the reduct, for DET-generation also the hierarchy-consistency is needed. In the classical case, only term-generation is considered, therefore sufficient completeness alone suffices for the corresponding theorem.

Theorem 2.32

Let $T = (\Sigma, R)$ be a DET-additive and DET-complete hierarchical specification over a deterministic basis containing the primitive part T0 $= (\Sigma 0, R0)$, $\Sigma 0 = (S0, F0)$.

If T is sufficiently complete and hierarchy-consistent, then T is hierarchy-faithful.

Proof:

Let $A0 \in DGen(T0)$. We construct the model A extending A0 analoguously to the construction $D\Sigma/R$. The carrier sets of the model A consist of a mixture between terms and values in the carriers of A0, replacing every primitive term by its value in A0. Formally, for a term $t \in W(\Sigma)$ with $T \vdash DET(t)$, we call this mixed term t^{A0}, defined by:

$t^{A0} =_{def} e$, where $I^A[t] = \{ e \}$ (e is unique, since $T \vdash DET(t)$);

$t^{A0} =_{def} f(t_1,...,t_n)^A$ iff $t = f(t_1^{A0},...,t_n^{A0}) \notin W(\Sigma 0)$.

The hierarchy-consistency ensures that $T \vdash DET(t) \Leftrightarrow T0 \vdash DET(t)$ for $t \in W(\Sigma 0)$, so T is "additive" with respect also to the deterministic $\Sigma 0$-terms, and we can use the construction used for $D\Sigma/R$, giving a model A such that (for $t \in W(\Sigma)$):

$I^A[t] = \{ [t'^{A0}] \mid t' \in W(\Sigma) \wedge T \vdash t \rightarrow t' \wedge T \vdash DET(t') \}$.

Consider now a term $t0 \in W(\Sigma 0)$. In this case, for every $t0' \in W(\Sigma)$ with $T \vdash DET(t0')$, sufficient completeness of T gives us a $t0'' \in W(\Sigma 0)$ such that $T \vdash DET(t0'')$ and $t0' \approx t0''$. Therefore $I^A[t0] = \{ [t0'^{A0}] \mid t0' \in W(\Sigma 0) \wedge T \vdash t0 \rightarrow t0' \wedge T \vdash DET(t0') \} = \{ [t0'^{A0}] \mid t0' \in W(\Sigma 0) \wedge T0 \vdash t0 \rightarrow t0' \wedge T0 \vdash DET(t0') \}$ (because of hierarchy-consistency). According to the definiton of $t0'^{A0}$, this means that $I^A[t0] \subseteq I^{A0}[t0]$. The reverse inclusion $I^A[t0] \supseteq I^{A0}[t0]$ is a consequence of the DET-generation of A0. ◊

In section 4.4, a class of specifications will be defined, for which sufficient completeness and hierarchy-consistency can be checked by rather simple syntactical criteria.

The last result of this section illustrates to which extent the preconditions of theorem 2.32 already determine the admitted models of a given specification. In fact, there is only one model (up to isomorphism), as long as the non-primitive part does not introduce any new sorts ("functional enrichment").

Theorem 2.33

Assume the preconditions of theorem 2.32, where T does not introduce new sorts, i.e. S = S0.

Let A0$\in$DGen(T0) be given, and let A' be an arbitrary model such that the Σ0-reduct of A' is A0.

Then A' is isomorphic to the model A constructed in the proof of theorem 2.32.

Proof:

Since S = S0, the reduct-condition means that the carrier sets of A' and A0 are identical. We show first that also in the carrier sets of A (according to theorem 2.32) only values from A0 appear.

For an arbitrary term t$\in$W(Σ), we have

$$I^A[t] = \{ [t'^{A0}] \mid t' \in W(\Sigma) \land T \vdash t \rightarrow t' \land T \vdash DET(t') \}.$$

Since S = S0, the term t is of primitive sort, and so is t'. Therefore, sufficient completeness gives for any t' a term t0'$\in$W(Σ0) such that t' $\approx$ t0'. Since t0'$^{A0}\in I^{A0}[t0']$, $I^A[t]$ contains only values from the carriers of A0.

It remains to show that the identity mapping is a homomorphism with respect to the operations in $\Sigma\backslash\Sigma$0. This follows from the chain of equivalences

$$e \in f^A(e_1,...,e_n) \quad (e, e_i \text{ in the carriers of A0})$$
$$\Leftrightarrow \quad e \in I^A[f(t_1,...,t_n)]$$

(with appropriate $t_i \in W(\Sigma$0)such that $I^A[t_i] = \{e_i\}$, due to term-generation).

Due to DET-generation, we can assume T $\vdash$ DET(t_i), and due to sufficient completeness (and S = S0) also $t_i \in W(\Sigma$0). Equivalences continued:

$$\Leftrightarrow \quad \exists\, t' \in W(\Sigma): T \vdash f(t_1,...,t_n) \rightarrow t' \land T \vdash DET(t') \land I^A[t'] = \{ e \}$$

Again we can assume t'$\in$W(Σ0), so e = t'^{A0}. Equivalences continued:

$$\Leftrightarrow \quad e \in f^{A'}(e_1,...,e_n)$$

(definition of the extension A' as in theorem 2.32.) $\Diamond$

Chapter 3

Structure of the Model Classes

This chapter is dedicated to a study of results concerning the relationship between various models of a nondeterministic specification. In particular, the notion of a Σ-homomorphism for multi-algebras is dealt with in the following sections.

The significance of this whole theoretical approach using homomorphisms and extremal models (initial and terminal ones) is estimated very differently by various researchers. It is obvious that any serious generalization of the classical notions of algebraic specifications has to address this topic, and this is the motivation for this chapter. However, readers may skip this whole chapter, if they are not interested in the material presented here.

In this chapter, the notion of a homomorphism for multi-algebras is defined. The presence of nondeterminism leads to the introduction of two different notions of homomorphism, which are used both in the theory of extremal models.

In a first pass, the general theory of multi-algebras is revisited from the structural point of view. A counterexample shows that in the general model class from above, an initial model does not always exist. A terminal model, however, can be constructed for every specification.

In a second pass, extremal models for specifications over a deterministic basis are investigated. It is shown that the term model $D\Sigma/R$, which was defined in the last chapter, is initial in some sense. A semantical characterization for the model class DGen(T) from the last chapter is given, and it is shown that the term model is initial within this model class in another, stronger sense.

3.1 Homomorphisms and Extremal Algebras

In order to compare two multi-algebras, the central notion is that of a *homomorphism*. A homomorphism can be established between two multi-algebras A and B, if B can be seen as an abstraction of A. An algebra B is here called an abstraction of A, if the structure of B can be completely described by the structure of A, where elements of the carriers of A are identified, possibly.

Definition 3.1 (Σ-Homomorphism)

Let $\Sigma = (S, F)$ be a signature, $A, B \in \text{MAlg}(\Sigma)$. A *(tight)* Σ– *homomorphism* φ from A to B is a family of mappings
$$\varphi = (\varphi_s)_{s \in S}, \qquad \varphi_s: s^A \to \wp^+(s^B),$$
which fulfils the following condition:

For all $[f: s_1 \times \ldots \times s_n \to s] \in F$ and all $e_1 \in s_1{}^A, \ldots, e_n \in s_n{}^A$:
$$\{ e' \in \varphi_s(e) \mid e \in f^A(e_1, \ldots, e_n) \}$$
$$= \{ e' \in f^B(e_1', \ldots, e_n') \mid e_1' \in \varphi_{s_1}(e_1), \ldots, e_n' \in \varphi_{s_n}(e_n) \}$$

φ is called a *loose* Σ-homomorphism, if the following, less restrictive, condition holds:
$$\{ e' \in \varphi_s(e) \mid e \in f^A(e_1, \ldots, e_n) \}$$
$$\subseteq \{ e' \in f^B(e_1', \ldots, e_n') \mid e_1' \in \varphi_{s_1}(e_1), \ldots, e_n' \in \varphi_{s_n}(e_n) \}$$

φ is called *element-valued*, iff for all $s \in S$: $\forall e \in s^A: |\varphi(e)| = 1$. $\Diamond$

The notion of a homomorphism, as it is defined above, is a bit more general than the definitions found in the literature. Homomorphisms for multi-algebras have been defined already in [Pickert 50] and later in [Pickett67], [Hansoul83], [Nipkow86] and [Hesselink88]. These papers always consider only *element-valued* homomorphisms instead of the set-valued definition from above. The definition above contains the element-valued homomorphism as a special case. The main reason why the generalization has been chosen is that it subsumes the interpretation of a term as a special case of a homomorphism. This question will be studied in more detail below.

All the definitions in literature contain a distinction between loose and tight homomorphisms. Unfortunately, the names vary from paper to paper. Loose homomorphisms enable sensible results, as it is shown below. A dual generalization (using "$\supseteq$" instead of "$\subseteq$") does not make any sense, since such a "homomorphism" can be always established between two arbitrary Σ-multi-algebras. (Simply choose $\varphi_s(e) = s^B$.)

Below a few examples for simple homomorphisms are given.

Example 3.2

Let $A \in MAlg(\Sigma)$, $W\Sigma$ the algebra of ground terms according to example 1.3. Then the interpretation mapping I^A

$$I^A: W\Sigma \to A, \quad I_s^A: W(\Sigma)_s \to \wp^+(s^A)$$

is a tight Σ-homomorphism:

$$\{e \in I^A[t] \mid t \in f^{W\Sigma}(t_1, \ldots, t_n)\}$$
$$= \{e \in I^A[t] \mid t \in \{ f(t_1,\ldots,t_n) \} \}$$
$$= I^A[f(t_1,\ldots,t_n)] \qquad \text{(Example 1.3)}$$
$$= \{e \in f^A(e_1,\ldots,e_n) \mid e_i \in I^A[t_i]\} \quad \text{(Definition 1.5)} \qquad \Diamond$$

Example 3.3

Let $A \in MAlg(\Sigma)$. The mapping $id = (id_s)_{s \in S}$
$$id_s: s^A \to s^A, \quad id_s(e) = \{ e \} \text{ für } e \in s^A$$
is a tight Σ-homomorphism:

$$\{e' \in id(e) \mid e \in f^A(e_1,\ldots,e_n)\}$$
$$= f^A(e_1,\ldots,e_n)$$
$$= \{e' \in f^A(e_1',\ldots,e_n') \mid e_i' \in id(e_i)\} \qquad \Diamond$$

Theorem 3.4

Let $A, B, C \in MAlg(\Sigma)$ and $\varphi_1: A \to B$, $\varphi_2: B \to C$ tight Σ-homomorphisms.

Then $\varphi_2 \cdot \varphi_1: A \to C$ is a tight Σ-homomorphism, again.

An analoguous result holds for loose Σ-homomorphisms.

Proof:

$$\{k \in (\varphi_2 \cdot \varphi_1)(e) \mid e \in f^A(e_1,\ldots,e_n)\}$$
$$= \{k \in \varphi_2(l) \mid l \in \varphi_1(e) \wedge e \in f^A(e_1,\ldots,e_n)\} \qquad \text{(Definition } \varphi_2 \cdot \varphi_1)$$
$$= \{k \in \varphi_2(l) \mid l \in f^B(l_1,\ldots,l_n) \wedge l_i \in \varphi_1(e_i)\}$$
$$\qquad\qquad\qquad\qquad\qquad\qquad\qquad (\varphi_1 \text{ is a homomorphism)}$$
$$= \{k \in f^C(k_1,\ldots,k_n) \mid k_i \in \varphi_2(l_i) \wedge l_i \in \varphi_1(e_i)\}$$
$$\qquad\qquad\qquad\qquad\qquad\qquad\qquad (\varphi_2 \text{ is a homomorphism)}$$
$$= \{k \in f^C(k_1,\ldots,k_n) \mid k_i \in (\varphi_2 \cdot \varphi_1)(e_i)\} \qquad \text{(Definition } \varphi_2 \cdot \varphi_1)$$

Analoguously for loose homomorphisms. $\Diamond$

Within the model class of a given specification, the most extreme models are of particular interest. These are the maximally refined and the maximally abstract model which are admitted by the specification.

Definition 3.5 (Initial and Terminal Algebra)

Let K be a class of Σ-algebras. An algebra $I \in K$ is called *(tightly) initial* in K, iff for every algebra $A \in K$ there exists exactly one Σ-homomorphism from I to A. $T \in K$ is called *terminal*, iff for every $A \in K$ there exists at least one (tight) element-valued Σ-homomorphism from A to T.

A is called *loosely initial*, iff the definition of initiality is fulfilled, where tight homomorphisms are replaced by loose ones. $\Diamond$

The definition of terminality above uses only element-valued homomorphisms; therefore it is consistent with the notions in the literature. For most of the initiality results, which are given below, it turns out that also only element-valued homomorphisms are involved.

Exactly like in the deterministic case, a rather trivial terminal algebra can be constructed easily:

Definition 3.6

For a given signature $\Sigma = (S, F)$, let an algebra $Z\Sigma$ be defined by
$$s^{Z\Sigma} = \{\, s \,\} \qquad \text{for } s \in S$$
$$f^{Z\Sigma}(s_1,\dots,s_n) = \{\, s \,\} \qquad \text{for } [f\colon s_1 \times \dots \times s_n \to s] \in F \qquad \Diamond$$

Theorem 3.7

$Z\Sigma$ is terminal in $Mod(T)$ for a given specification $T = (\Sigma, R)$.

Proof:

By induction on the term structure of t for an arbitrary valuation β in $Z\Sigma$ the following fact can be shown:
$$\forall\ t \in W(\Sigma)_s\colon I_\beta^{Z\Sigma}[t] = \{\, s \,\} \,.$$

Therefore, for an axiom $\langle l \to r \rangle \in R$, where l and r are of the same sort:
$$I_\beta^{Z\Sigma}[l] = \{\, s \,\} = I_\beta^{Z\Sigma}[r].$$

This means that $Z\Sigma \in Mod(T)$.

For $A \in Mod(T)$ the mapping
$$\varphi\colon A \to Z\Sigma, \qquad \varphi_s(e) = \{\, s \,\} \qquad \text{für } e \in s^A$$
is a tight Σ-homomorphism:
$$\{e' \in \varphi_s(e) \mid e \in f^A(e_1,\dots,e_n)\} = \{\, s \,\}$$
$$= \quad \{e' \in f^{Z\Sigma}(e_1,\dots,e_n) \mid e_i' \in \varphi_{s_i}(e_i)\} \qquad\qquad \Diamond$$

The fact that there is a unique (up to isomorphism) terminal algebra is due to the definition of terminality which refers to element-valued homomorphisms only. If the notion of a terminal algebra was formulated with abitrary (set-valued) homomorphisms, an infinite number of non-isomorphic terminal models would be admitted. For a similar reason, "loosely terminal" models are not studied here.

The algebra $W\Sigma$ of ground terms (from example 1.3) can be shown to be initial within all multi-algebras of a given signature, as in the classical case. This is only possible since the notion of non-element-valued homomorphisms has been introduced here.

Theorem 3.8

WΣ is tightly initial in MAlg(Σ).

Proof:

The existence of a homomorphism φ: WΣ $\rightarrow$ A can be shown analoguously to example 1.22, its uniqueness can be shown by induction on the term structure, like in [ADJ 78]. $\Diamond$

3.2 Initial Models

The result above was about initiality in the general class of all algebras of a given signature. This section now addresses the question of initiality within the class of all models of a nondeterministic specification. In a first approach, the general notion of a nondeterministic specification is presupposed, as it was used in chapter 1. Please note that this means in some sense a step backwards compared with the material of chapter 2! In order to keep this exposition as short as possible, we restrict ourselves here to the simplest case of *ground specifications*, where the axioms do not contain free variables. Despite of this restriction, it can be shown that also in general initial algebras do *not* exist.

For this result, the notion of a *term-generated* model, as it was defined in definition 2.21, is needed again. A model A of a specification is called *term-generated*, if for every object e in the algebra there is a ground term t which describes the object: $I^A[t] = \{ e \}$.

Theorem 3.9

Let T = (Σ, R) be a ground nondeterministic specification.
If a multi-algebra C$\in$Mod(T) is loosely initial in Mod(T), then C is term-generated (i.e. C$\in$Gen(Σ)).

Proof:

For a ground specification T, it is easy to construct a ground term model W which fulfils for every ground term t$\in$W(Σ):
$$I^W[t] = \{t' \in W(\Sigma) \mid T \vdash_{RC} t \rightarrow t' \}$$

(The details of this construction are as in theorem 1.19, but for ground terms only.)

Since C is loosely initial, there is a homomorphism ψ from C to the ground term model W.

The interpretation I^C of ground terms gives a homomorphism from the algebra W to the algebra C. This is due to the fact that the rewriting calculus is sound for ground specifications (theorem 1.17), which means that

$$\{e \in I^C[t] \mid T \vdash_{RC} f(t_1,...,t_n) \to t\} \subseteq I^C[f(t_1,...,t_n)].$$

Now $I^C \cdot \psi$: C→C is a homomorphism from C to C (theorem 3.4)

According to example 3.3, another homomorphism from C to C is given by the identity (id). Initiality means that the homomorphism from C to C is unique, therefore for a given $e \in s^C$ holds

$$(I^C \cdot \psi)(e) = \{e\}, \text{ i.e. } \{e' \in I^C[t] \mid t \in \psi(x)\} \subseteq \{e\}.$$

This means $I^C[t]=\{e\}$ for all $t \in \psi(e)$. Since $\psi(e) \neq \emptyset$, there is a $t \in W(\Sigma)$ such that $I^C[t] = \{e\}$.   ◊

The following example is used for the demonstration that in Mod(T) loosely initial algebras do not always exist.

Example 3.10

```
spec NI
sort s
func    a: → s,          b: → s,
        g: → s,          f: s → s
axioms
        g → a,           f(a) → b,
        f(b) → a,        f(g) → a
end
```

Two non-isomorphic models A and B for NI are defined by:

$s^A = s^B = \{ a, b \}$,

$a^A = a^B = \{ a \}$, $b^A = b^B = \{ b \}$,

$g^A = \{ a \}$, $g^B = \{ a, b \}$,

$f^A(a) = \{ a, b \}$, $f^B(a) = \{ b \}$,

$f^A(b) = \{ a \}$, $f^B(b) = \{ a \}$.   ◊

The specification NI is similar to INC from example 2.7 (which was used as a counterexample for general incompleteness of rewriting). It has been chosen in such a way that in all models the inclusion

$$f(g) \to a$$

has to hold. But it is in no way clear whether this is a restriction which applies to the functions f or g. In the first case, a more precise axiom is

$$f(a) \to a$$

the latter case can be described also by

$$g \to b .$$

The models A and B realise these both choices. These choices cannot be both represented within a (term-generated) initial model.

Theorem 3.11

Let $T = (\Sigma, R)$ be a nondeterministic specification.

In general, in Mod(T) loosely initial multi-algebras do not exist.

Proof:

Consider the class Mod(NI) of all models of NI, as it was defined in example 3.10, together with the two models A and B. Without loss of generality, let $a \neq b$ for the elements of the carrier sets of A and B.

Let C be a loosely initial algebra in this class. Initiality means that there are homomorphisms

$$\varphi_A: C \to A \quad \text{and} \quad \varphi_B: C \to B.$$

From the homomorphism condition for φ_A, applied to the functions a and b, the following propositions follow:

(1) $\varphi_A(e) = \{ a \}$ for all $e \in a^C$, $\varphi_B(e) = \{ a \}$ for all $e \in a^C$

(2) $\varphi_A(e) = \{ b \}$ for all $e \in b^C$, $\varphi_B(e) = \{ b \}$ for all $e \in b^C$

From the homomorphism condition for φ_A and function g follows:

(3) $\{ e' \in \varphi_A(e) \mid e \in g^C \} \subseteq g^A = \{ a \}$

Let $e_b \in b^C$. Assume that $e_b \in g^C$; then from (3) and (2) follows $\{ b \} \subseteq \{ a \}$, which contradicts to $a \neq b$. Therefore:

(4) $\forall e \in b^C: e \notin g^C$

Let $e_a \in a^C$. From the homomorphism condition for φ_B and function f follows:

(5) $\{ e' \in \varphi_B(e) \mid e \in f^C(e_a) \} \subseteq \{ e' \in f^B(e_1) \mid e_1 \in \varphi_B(e_a) \}$

Using (1), this means:

(6) $\{ e' \in \varphi_B(e) \mid e \in f^C(e_a) \} \subseteq \{ b \}$

Assume now that $e_a \in f^C(e_a)$; from (6) and (1) follows $\{ a \} \subseteq \{ b \}$, which contradicts to $a \neq b$. Therefore:

(7) $\forall\, e \in a^C$: $e \notin f^C(e)$

In model C, the inclusion $\langle f(g) \to a \rangle$ has to hold. Therefore:

(8) $\exists\, e_a \in a^C, e_0 \in g^C$: $e_a \in f^C(e_0)$

Since model C is term-generated (theorem 3.9), there must be a ground term t_0 the interpretation of which is e_0: $I^C[t_0] = \{ e_0 \}$. The axioms of NI ensure that every ground term can be reduced either to a or b. Applied to t_0, this means that $\{ e_0 \} \supseteq a^C$ or $\{ e_0 \} \supseteq b^C$. Slightly rephrased, this is:

(9) $e_0 \in a^C \ \vee \ e_0 \in b^C$

From (7) and (8) follows that $e_0 \notin a^C$. From (4) and (8) follows that $e_0 \notin b^C$. This is a contradiction to (9).

To summarize, an appropriate algebra C does not exist. $\Diamond$

This concludes the discussion of initial models for the general case. The theorem above may be seen as an additional argument why the extension to specifications on a deterministic basis, as it has been introduced in chapter 2, is useful. So let us now turn to the case of specifications with a deterministic basis.

3.3 Initial Models with Deterministic Basis

The aim of this section is to show that the term model DΣ/R for DET-complete and DET-additive specifications, as it has been defined in definition 2.13, is an initial model. This model is term-generated. Therefore, the following lemma is useful which states the consequences of these conditions onto homomorphisms from the initial model to an arbitrary model. As far as DET-complete specifications and term-generated models are concerned, the set-valued notion of homomorphism coincides with the classical notion of homomorphism. As long as initial models in Mod(T) are term-generated, as it is suggested by theorem 3.9, this shows the consistency between the notion of homomorphism as it is used here and the literature on homomorphisms and initiality.

Lemma 3.12

Let T be a DET-complete specification, A∈Gen(T), B∈Mod(T). Then every loose homomorphism $\varphi: A \to B$ is element-valued (i.e. it assigns only singleton sets).

Proof:

Let $e \in s^A$. Since A is term-generated, there is a $t \in W(\Sigma)$ such that $I^A[t] = \{e\}$. Since T is DET-complete, there is a t' such that |- DET(t') and |- t→t', so $I^A[t'] = \{e\}$. The definition of a homomorphism gives $\varphi(e) \subseteq I^B[t']$. Because of |- DET(t') we have $|\varphi(e)| \le 1$.   ◊

Now the expected initiality result for DΣ/R can be shown.

Theorem 3.13

Given a DET-complete and DET-additive specification $T = (\Sigma, R)$, DΣ/R is loosely initial in Mod(T).

Proof:

Let A∈Mod(T). Define the mapping
$$\varphi: D\Sigma/R \to A$$
as the extension of the interpretation I^A to the carriers of DΣ/R :
$$\varphi([t]) = I^A[t] \qquad \text{where } t \in_s D\Sigma/R.$$
The well-definedness of φ is a consequence of theorem 2.6 and definition 2.12.

For the remaining parts of the proof see appendix A.   ◊

DΣ/R is loosely initial in Mod(Σ), but not tightly initial. This is demonstrated by the following example.

Example 3.14

spec NPI
sort s
func   a: → s,   b: → s,   g: → s,   f: s → s
axioms
          DET(a),   DET(b),   g → a,   f(x) → x
end

A model A of NPI is given by
$$s^A = \{\, a, b \,\}, \quad a^A = \{\, a \,\}, \quad b^A = \{\, b \,\}, \quad g^A = \{\, a\,, b \,\},$$
$$f^A(e) = \{\, e \,\} \quad \text{for } e, e1, e2 \in \{a, b \,\}.$$

Within DΣ/NPI, the loosely initial model, we have:
$$g^{D\Sigma/NPI} = \{\, [a] \,\}.$$

Therefore, the condition of a loose homomorphism
$$\{ e \in \varphi([t]) \mid [t] \in g^{D\Sigma/NPI}\} = \{\, a \,\} \subseteq \{\, a, b \,\} = g^A$$
holds, but not the condition of a tight homomorphism (which involves set equality instead of subset relation). $\Diamond$

Obviously, the model class has to be restricted, in order to show a tight initiality result. Example 3.14 shows also that the restriction to term-generated models is not sufficient for this purpose. A good candidate for an appropriate model class is the class DGen(T) of term-generated and DET-generated models, as it has been introduced in the last chapter (definition 2.26). (As a reminder: The models in DGen(T) are those where the interpretation of a nondeterministic term contains only elements which can be reached by a deterministic term, and where this inclusion can be derived on the level of terms within the calculus.)

Before showing an initiality result, we address the general question of how to characterize this model class DGen(T). It turns out that the notion of homomorphism can be used to give a more "semantic" characterization, which does not involve any reference to deduction.

The basic idea is the observation that the models in DGen(T) are "maximally deterministic" in the sense that they do not contain any "superfluous" non-determinism which is not explicitly mentioned in the specification. In order to speak about degrees of determinacy, the notion of a "descendant" (analoguously to [McCarthy 61]) is used.

Definition 3.15 (Descendant)

Let $T = (\Sigma, R)$ be a specification, $A \in Gen(T)$. Another model $A' \in Gen(T)$ is called a *descendant* of A, iff:
$$\forall\, t \in W(\Sigma): I^A[t] \supseteq I^{A'}[t].$$

A' is called a *proper descendant* of A, iff A' is a descendant of A and if
the additional condition holds:

$$\exists\ t \in W(\Sigma): I^A[t] \neq I^{A'}[t]. \qquad\qquad \Diamond$$

If an algebra has proper descendants, it must not be called maximally
deterministic. Unfortunately, this does not suffice to characterize maximally
deterministic algebras. There are more complex cases of "superfluous"
nondeterminism, as the following example shows.

Example 3.16

There is a proper descendant of the algebra A from example 3.14 above:

$$s^{A'} = \{\, a\, , b\, \},$$
$$a^{A'} = \{\, a\, \}, \qquad b^{A'} = \{\, b\, \}, \qquad g^{A'} = \{\, a\, \},$$
$$f^{A'}(e) = \{\, e\, \} \qquad \text{where } e, e1, e2 \in \{\, a, b\, \}.$$

There is a loose homomorphism $\varphi: A' \to A$, defined by

$$\varphi(a) = \{\, a\, \}, \qquad \varphi(b) = \{\, b\, \}\, .$$

The new algebra A' does not have any proper descendants.

A more complex case is the following one:

spec NMD
sort   s
func   a: → s,   b: → s,   c: → s,   f: s → s
axioms
         DET(a),   DET(b),   DET(c),
         f(a) → a,   f(b) → c,   f(c) → c
end

with the model B:

$$s^B = \{\, a\, , c\, \},$$
$$a^B = \{\, a\, \}, \qquad\qquad\qquad b^B = \{\, a\, \}, \qquad\qquad\qquad c^B = \{\, c\, \},$$
$$f^B(x) = \{\, a, c\, \}, \quad f^B(c) = \{\, c\, \}\, .$$

B does not have any proper descendants. But if B is "refined" (extending
its carrier set), a "less deterministic" model can be constructed, which is
called B':

$$s^{B'} = \{\ a1\ , a2\ , c\ \},$$
$$a^{B'} = \{\ a1\ \}, \qquad b^{B'} = \{\ a2\ \}, \qquad c^{B'} = \{\ c\ \},$$
$$f^{B'}(a1) = \{\ a1\ \}, \quad f^{B'}(a2) = \{\ c\ \}, \quad f^{B'}(c) = \{\ c\ \}.$$

Again there is a loose homomorphism ψ: B' → B:

$$\psi(a1) = \{\ a\ \}, \qquad \psi(a2) = \{\ a\ \}, \qquad \psi(c) = \{\ c\ \}. \qquad\qquad \Diamond$$

The term model DΣ/NMD gives a deterministic interpretation for the operation f, therefore a maximally deterministic model should interpret f also as deterministic. The example gives a hint, how this property can be formulated in terms of models: An algebra A is maximally deterministic iff it does not have a more deterministic refinement:

Definition 3.17 (Maximally Deterministic)

Let A, A' be term-generated Σ-algebras.

A' is called a *refinement* of A, iff there is a loose Σ–homomorphism φ: A' → A.

A' is called *more deterministic* than A, iff:

$$\forall\ t \in W(\Sigma): \ |\ I^A[t]\ | \geq |\ I^{A'}[t]\ |.$$

A is called *maximally deterministic*, iff A is more deterministic than every refinement of A. $\qquad\qquad \Diamond$

The next lemma shows that the semantic characterization of maximal determinacy coincides with the model class DGen(T). Moreover, it shows a useful property about homomorphisms, which leads to the immediate consequence that a loosely initial model in DGen(T) is also a tightly initial one.

Lemma 3.18

Let $T = (\Sigma, R)$ be a DET-complete and DET-additive specification, A∈Gen(T). Then the following three propositions are equivalent:

(1)   A is maximally deterministic.

(2) $\forall\ B \in$ Gen(T):

φ: B→A is a loose Σ–homomorphism $\Rightarrow$
φ is a tight Σ-homomorphism.

(3)   A $\in$ DGen(T).

Proof: See appendix A.   ◊

Please note that lemma 3.18 assumes the specification to be DET-additive. The
results cannot be generalized easily to non-additive specifications.

A consequence of lemma 3.18 (3) is the fact that DΣ/R is maximally deterministic. From this fact an initiality result follows:

Theorem 3.19

> Let T = (Σ, R) be a DET-complete and DET-additive specification.
> Then DΣ/R is tightly initial in DGen(T).

Proof:

> Consequence of theorem 3.13 and lemma 3.18 (2), since
> DΣ/R$\in$DGen(T).   ◊

We conclude this chapter with a graphical sketch of the lattice structure
connecting the models of a specification T = (Σ, R).

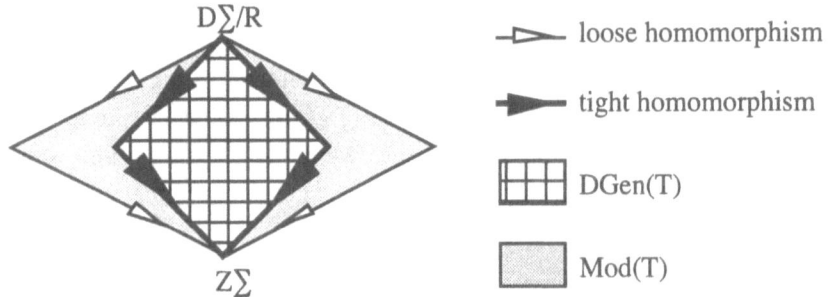

Chapter 4

Nondeterministic Specifications as a General Framework

At this point, the presented theory has reached a stage, where it is interesting to investigate the relationship to classical concepts like the theory of term rewriting and equational logic as well as logic programming. It will turn out that equational logic and (confluent) term rewriting can be seen as special cases of the new theory. It also will be shown that a special variant of the theory has very close connections to algebraic and logic programming.

The general observation is that nondeterministic specifications form a rather general framework which is well-suited for integrating and comparing various approaches from denotational and operational semantics. Even beyond the examples mentioned above, in [Meseguer 92] a whole catalogue of concepts from computer science can be found, which can be subsumed by a variant of rewriting, if the confluence restrictions are left out.

4.1 Equational Logic

Equational logic can be easily integrated into our new framework. The basic idea is that a given equation is simulated by two rewrite rules, which differ only in exchanged left and right hand sides. This way, the symmetry deduction rule can be simulated within term rewriting.

In the framework of specifications over a deterministic basis, this effect can be achieved in an even simpler way.

Definition   4.1

>For a given equational specification
>$$T = (\Sigma, E), \qquad \Sigma = (S, F),$$
>a nondeterministic specification NDEQ(T)
>$$NDEQ(T) = (\Sigma, R)$$
>can be constructed using:
>$$R = \{ \: l \rightarrow r \: | \: l = r \in E \} \cup \{ \: \langle DET(f(x_1, \ldots, x_n) \rangle \: | \: f \in F \: \}$$
>(where $x_1, \ldots, x_n$ are pairwise disjoint variables). $\Diamond$

The DET-axioms are chosen in such a way that all operations are deterministic. The simulation of the symmetry deduction rule is now achieved by the deduction rule (DET-R).

Theorem 4.2 (Simulation of Equational Reasoning)

>Within the nondeterministic specification NDEQ(T) = (Σ, R) associated to an equational specification T, the following holds:
>(1) $\forall \: t \in W(\Sigma, X)$: R $\vdash$ DET(t)
>(2) $\forall \: t1, t2 \in W(\Sigma, X)$: t1 $=_E$ t2 $\Leftrightarrow$ R $\vdash$ t1 $\rightarrow$ t2.

Proof:

>(1) can be easily shown by structural induction on the number of function symbols in t. Either there is no function symbol in t (then we can use (DET-X)), or we can apply (AXIOM-2) using one of the DET-axioms contained in R.

>(2) is shown by induction on the length of the derivation for t1 $=_E$ t2 within the classical equational calculus. The cases of reflexivity, transitivity and congruence (with respect to term building operations) can be directly covered using (REFL), (TRANS), (CONG). The remaining cases are:
>Application of an equation:
>>Here t1 = σl, t2 = σr, $l = r \in E$. Due to (1), we have
>>$\forall \: x \in X$: R $\vdash$ DET(σx).
>>Therefore, (AXIOM-1) can be applied to show R $\vdash$ t1 $\rightarrow$ t2.

Symmetry:

> Here t2 $=_E$ t1. By induction hypothesis, R |- t2 → t1.
> Due to (1), we have R |- DET(t2), so (DET-R) gives:
> R |- t2 → t1.   ◊

In other words, the calculus of definition 2.4 in this case exactly agrees with the equational calculus. Morever, theorem 4.2 (1) ensures that the interpretation of every term in any model of NDEQ(T) is a singleton set. Therefore, all models of NDEQ(T) correspond exactly to classical Σ-algebras.

4.2 Term Rewriting

In difference to the above results on equational logic, classical term rewriting cannot be subsumed by the rewriting relation as it has been axiomatized in definition 2.4. It was one of the main results of section 1 that classical term rewriting is unsound for the semantic framework of heterogeneuous multi-algebras. This is also the main point where the approach studied here differs from the work of Meseguer ([Meseguer 92]).

However, one would expect confluent axiom systems to show some particular semantical properties. Please note that the notion of confluence here refers to the rewriting relation between terms as it is established by the calculus of deinition 2.4. Moreover, we restrict our attention to so-called ground confluence. A set of inclusion rules is called *ground confluent*, iff the rewriting relation generated by the calculus of definition 2.4 is confluent on ground terms.

Theorem 4.3

> If R is ground confluent, then in A∈DGen(Σ, R) all operations are deterministic.

Proof:

> Let e1, e2∈I^A[t], t∈W(Σ). Since A is DET-generated (see definition 2.21), there are t1, t2∈W(Σ) such that
>> |- DET(t1), |- DET(t2), |- t → t1, |- t → t2,
>> I^A[t1] = { e1 }, I^A[t2] = { e2 }.
>
> Ground confluence ensures that there is a t' such that

$\vdash t1 \to t'$, $\vdash t2 \to t'$.

Using (DET-R), we have $\vdash t' \to t2$, using (TRANS) $\vdash t1 \to t2$. Therefore, using theorem 2.6: $\{ e1 \} \supseteq \{e2\}$, i.e. $e1 = e2$. Hence

$$\mid I^A[t] \mid = 1 .$$

$\Diamond$

Ground confluence forces all terms to be deterministic. Using the breadth-induction calculus from definition 2.18, we can even deduce for every ground term t the formula DET(t). A difference to equational deduction is that there determinacy is ensured for all models and even for non-ground terms. In the case of non-confluent (nondeterministic) rewriting, the more refined notion of the model class DGen(T) is used, and only for these models and for ground terms the determinacy is enforced. These observations correspond closely to the various levels of abstraction described by [Meseguer 92] (for a summary see Fig. 5 in [Meseguer 92]).

Another nice property of ground confluence is that it automatically ensures DET-additivity.

Theorem 4.4

> If R is ground confluent and DET-complete, then R is DET-additive, too.

Proof:

> Let $\vdash f(t_1,\ldots,t_n) \to t$, $\vdash$ DET(t). Because of DET-completeness, there are t_1', ..., t_n' where $\vdash$ DET(t_i'), $\vdash t_i \to t_i'$. With (CONG) and (TRANS): $\vdash f(t_1,\ldots,t_n) \to f(t_1',\ldots,t_n')$. According to ground confluence, there is a t' such that $\vdash f(t_1',\ldots,t_n') \to t'$, $\vdash t \to t'$. Using (DET-R), it follows that $\vdash t' \to t$, therefore (using (TRANS)) $\vdash f(t_1',\ldots,t_n') \to t$.

$\Diamond$

In many cases, also DET-completeness can be guaranteed automatically. For this purpose, it is necessary that every term has a normal form with respect to $\to$ and that the $\to$-terminal terms can be enumerated. Then a ground confluent set R of term rewrite rules over the signature Σ is transformed into the nondeterministic specification

$$T = (\Sigma, R \cup \{\langle DET(t)\rangle \mid t \text{ is } \to\text{-terminal} \}).$$

According to theorem 4.3, the DET-axioms hold within DGen(Σ,R). This means that they can be added without changing the semantics. The DET-axioms ensure the DET-completeness and theorem 4.4 gives the DET-additivity. In this case, the semantics given by DGen corresponds exactly to equational logic. The advantage of the confluent rewriting system is that there are less DET-axioms needed, and that the automatic search of deductions is considerably easier, due to the fact that the rule (DET-R) is avoided. As in the classical case, deduction here can be restricted to uni-directional application of the axioms.

Another aspect of non-confluent term rewriting can be quite interesting in some applications, where the confluence of a term rewriting system is yet unknown. The ideas from above give a semantics for term rewriting, independently of the confluence of the axiom system. If ground confluence (for the rewriting relation from definition 2.4) holds, this semantics automatically coincides with the usual semantics.

4.3 Conditional Axioms

A generalization of algebraic specifications to conditional axioms is interesting, mostly for the reason that here the central results still hold and the correspondence to term rewriting and equational logic is kept. The results for equational logic can be carried over to conditional-equational axioms (see for instance [Broy, Wirsing 82]); *conditional term rewriting systems* ([Kaplan 84], [Bergstra, Klop 86]) give an operational semantics for such specifications with conditional axioms. Below follows a sketch of the way how conditional axioms can be integrated into the framework presented here.

It is quite obvious how the syntax and semantics of conditional inclusion rules is to be defined. This differs from the situation in conditional term rewriting, where at least three variants of conditional axioms are distinguished. The three variants correspond to the following schemes of axioms:

$$\begin{array}{lll} \text{(a)} & t1 \leftrightarrow_R^* t2 & \Rightarrow \quad l \rightarrow_R r \\ \text{(b)} & t1 \downarrow_R t2 & \Rightarrow \quad l \rightarrow_R r \\ \text{(c)} & t1 \rightarrow_R^* t2 & \Rightarrow \quad l \rightarrow_R r \end{array}$$

Variant (a) admits conditions of the form "t1 is equivalent to t2", which can be proven by arbitrary applications of the axioms (including "backward" applications). In variant (b), the condition can be only fulfilled if both terms (t1 and t2) have a common successor within the term rewriting relation ($\rightarrow_R^*$). Variant (c) ist the simplest one: Conditions can be fulfilled there only if an (oriented) rewriting relation between the two terms can be proven.

For non-confluent rewriting, variant (b) is not very interesting, since the relation t1 $\downarrow_R$ t2 carries useful information only in the case of confluent rewriting (then it is equivalent to t1 $\leftrightarrow_R^*$ t2).

For similar reasons, variant (a) can be excluded. The relation t1 $\leftrightarrow_R^*$ t2 is useful for non-confluent specifications (it was called $\approx$ above). But, if only conditions of this kind were admitted, the language would be more restrictive than necessary. In general, the most interesting type of conditions is (c), where the application of a certain axiom is dependent of the question whether a term t1 can be brought into the shape of t2 (by nondeterministic rewriting). Note that axioms of the type (b) can be simulated within this approach by

$$\text{t1} \rightarrow \text{t2} \ \& \ \text{t2} \rightarrow \text{t1} \quad \Rightarrow \text{l} \rightarrow \text{r.}$$

Definition 4.5 (Conditional Inclusion Rule)

A *conditional (Σ, X-) inclusion rule* is a pair, consisting of a finite sequence of Σ, X-inclusion rules (the *condition*) and an atomic Σ, X-inclusion rule (*conclusion*). In formula notation this is written:

$$\text{t}_1 \rightarrow \text{t}_1' \ \& \ ... \ \& \ \text{t}_n \rightarrow \text{t}_n' \Rightarrow \text{l} \rightarrow \text{r}$$

where for i$\in\{1,...,n\}$: t_i, $t_i' \in W(\Sigma, X)_{s_i}$, $s_i \in S$, $l, r \in W(\Sigma, X)_s$, $s \in S$. $\Diamond$

Definition 4.6 (Validity for Conditional Rules)

Let A be a Σ-multi-algebra. A conditional inclusion rule is called *valid* in A, symbolically

$$A \models \text{t}_1 \rightarrow \text{t}_1' \ \& \ ... \ \& \ \text{t}_n \rightarrow \text{t}_n' \Rightarrow \text{l} \rightarrow \text{r,}$$

iff the following proposition holds:

Every valuation $\beta \in$ ENV(X,A) which fulfils the following condition:

$$\forall \ i\in\{1,...,n\}: \ I_\beta^A[t_i] \supseteq I_\beta^A[t_i'],$$

also fulfils

$$I_\beta^A[l] \supseteq I_\beta^A[r]. \hspace{3cm} \Diamond$$

The notion of a model is obvious. The next definition defines a suitable calculus for nondeterministic specifications with conditional axioms.

Definition 4.7 (Term Rewriting Calculus with DET and Conditional Rules)

Let $T = (\Sigma, R)$ be a nondeterministic specification containing DET-axioms and conditional inclusion rules. Then a formula ‹t1→t2› or ‹DET(t)›, respectively, is called *deducible* in T, written

$$T \vdash_{\text{COND}} t1 \to t2 \quad \text{or} \quad T \vdash_{\text{COND}} DET(t), \text{ respectively,}$$

iff there is a derivation for this formula using the following deduction rules:

(REFL), (TRANS), (CONG), (AXIOM-2),
(DET-X), (DET-D), (DET-R)   (as in definition 2.4)

(AXIOM-1-COND)

$$\frac{DET(\sigma x_1), ..., DET(\sigma x_n), \sigma t_1 \to \sigma t_1', ..., \sigma t_n \to \sigma t_n'}{\sigma l \to \sigma r}$$

if ‹t$_1$→t$_1$'› & ... & t$_n$→t$_n$' ⇒ l→ r ∈ R, σ∈SUBST(Σ, X),
{x$_1$,...,x$_m$} =Vars(l)∪Vars(r)∪Vars(t$_1$)∪...∪Vars(t$_n$)∪
                                    Vars(t$_1$')∪...∪Vars(t$_n$')

(AXIOM-1) is now a special case of (AXIOM-1-COND).   ◊

The following theorem shows the soundness of this calculus.

Theorem 4.8 (Soundness)

Using the preconditions of definition 4.7, for t, t1, t2 $\in$ W(Σ, X) the following implications hold:

$$T \vdash_{\text{COND}} t1 \to t2 \quad \Rightarrow \quad Mod(T) \models t1 \to t2$$
$$T \vdash_{\text{COND}} DET(t) \quad \Rightarrow \quad Mod(T) \models DET(t)$$

Proof: By induction on the length of the derivation, see appendix A.   ◊

The example below illustrates a new problem which arises now for the completeness of the calculus.

Example 4.9

> **spec** CIC
> **sort**   s
> **func**   a: $\rightarrow$ s, b: $\rightarrow$ s,
>   f: $\rightarrow$ s, g: $\rightarrow$ s
> **axioms**
>   DET(a),   DET(b),
>   f $\rightarrow$ a, f $\rightarrow$ b, g $\rightarrow$ a,
>   f $\rightarrow$ g $\Rightarrow$ a $\rightarrow$ b
> **end**

In DGen(CIC), independently of the condition, the following inclusion is valid (see the breadth-induction rule (IND-R)):
>   DGen(CIC) $\models$ f $\rightarrow$ g .

Therefore (using the conditional axiom):
>   DGen(CIC) $\models$ a $\rightarrow$ b .

But, according to definition 4.7:
> $\neg$(CIC $\vdash_{COND}$ f $\rightarrow$ g) and $\neg$(CIC $\vdash_{COND}$ a $\rightarrow$ b) $\Diamond$

The example shows that the calculus for conditional axioms is incomplete, even if, like in the unconditional case, only derivations for formulas of the shape
> $\vdash_{COND}$ t $\rightarrow$ t'   where $\vdash_{COND}$ DET(t')

are considered. Conditional deductions of such formulas can lead recursively to the deduction of other formulas which do not have the special shape. There are two ways to obtain a completeness result:

(1)   The calculus can be augmented by the breadth-induction rules (IND-R) and (IND-D). In this case, a rather complex calculus is the result. The deduction rule (AXIOM-1-COND) contains a premise which may lead to an inductive proof which in turn makes use of conditional deductions. In the unconditional case (definition 2.18), this kind of mutual recursion could be excluded.

(2)   The conditional axioms can be restricted syntactically. A simple case is achieved, if the preconditions of all conditional axioms have the shape
>   t $\rightarrow$ t'   where $\vdash_{COND}$ DET(t') .

Then the arguments of the unconditional case can be carried over to the conditional case.

The second variant is technically easier and it is sufficient for an interesting range of applications. Therefore it is studied in more detail.

Definition 4.10 (Simple Conditional Rules)

A conditional Σ, X-inclusion rule
$$\langle\, t_1 \rightarrow t_1' \ \& \ ... \ \& \ t_n \rightarrow t_n' \ \Rightarrow \ l \rightarrow r \,\rangle \in R$$
is called *simple*, iff:
$$\forall\, i \in \{1,...,n\}: T \vdash\text{-COND } DET(t_i'). \qquad\qquad \Diamond$$

The notions of DET-completeness and DET-additivity are extended in analogy to definition 2.8 / 2.9 (using the new calculus, $\vdash$-COND).

For DET-complete and DET-additive specifications containing only simple rules, again an initial model can be constructed.

Theorem 4.11

Let $T = (\Sigma, R)$ be a DET-complete and DET-additive specification which contains only simple conditional rules. Analoguously to definition 2.12, a model $D\Sigma/R$ is constructed, where for $t \in W(\Sigma)$ the interpretation is given by:
$$I^{D\Sigma/R}[t] = \{\, [t'] \mid T \vdash\text{-COND } DET(t') \wedge T \vdash\text{-COND } t \rightarrow t' \,\}.$$
$D\Sigma/R$ is initial in MGen(T), and for $t1$, $t2 \in W(\Sigma)$ the proposition holds:
$$D\Sigma/R \models t1 \rightarrow t2 \ \Leftrightarrow \ MGen(T) \models t1 \rightarrow t2 .$$

Proof: See appendix A. $\qquad\qquad \Diamond$

Conditional rewriting leads to a number of interesting theoretical problems even in the classical case. A detailed explanation of the underlying theory for the classical case has been given for instance in [Wechler 91].

A final example demonstrates the practical use of conditional axioms:

Example 4.12

A standard example from logic programming is the splitting of a sequence of data objects. The given sequence is splitted into two parts the concatenation of which results in the given sequence again.

Obviously, there is a (nondeterministic) choice, where to split the sequence. A corresponding specification is:

spec SPLIT
sort   Elem, Seq, Pair
func   { Operations for the sort Elem are omitted here }
            empty: $\rightarrow$ Seq
            append: Seq $\times$ Elem $\rightarrow$ Seq
            conc: Seq $\times$ Seq $\rightarrow$ Seq
            pair: Seq $\times$ Seq $\rightarrow$ Pair   { Pairs of sequences }
            split: Seq $\rightarrow$ Pair
axioms
            DET(empty),   DET(append(s,x)),   DET(pair(s1,s2)),
            conc(s,empty) $\rightarrow$ s,
            conc(append(s1,x),s2) $\rightarrow$ append(conc(s1,s2),x),
            conc(s1,s2) $\rightarrow$ s $\Rightarrow$ split(s) $\rightarrow$ pair(s1,s2)
end

SPLIT is DET-complete and DET-additive, the conditional axiom in SPLIT is simple.

Below follows a deduction for
  split(append(append(empty,b),a))
          $\rightarrow$ pair(append(empty,a),append(empty,b)),
where a,b $\in$ Elem such that |-DET(a), |-DET(b):
(1) |-COND DET(empty)   (AXIOM-2)
(2) |-COND DET(append(empty,a))   (AXIOM-2), (1)
(3) |-COND DET(append(empty,b))   (AXIOM-2), (1)
(4) |-COND DET(append(append(empty,b),a))   (AXIOM-2), (3)
(5) |-COND conc(empty,append(empty,b)) $\rightarrow$ append(empty,b)
                                                   (AXIOM-1), (3)
(6) |-COND append(conc(empty,append(empty,b)),a)
                $\rightarrow$ append(append(empty,b),a)   (CONG), (5)
(7) |-COND conc(append(empty,a),append(empty,b))
                $\rightarrow$ append(conc(empty,append(empty,b)),a)
                                                   (AXIOM-1), (1),(3)
(8) |-COND conc(append(empty,a),append(empty,b))
                $\rightarrow$ append(append(empty,b),a)   (TRANS), (6), (7)

(9) ⊢-COND split(append(append(empty,b),a))
  → pair(append(empty,a),append(empty,b))
    (AXIOM-1-COND), (4), (2), (3), (8)   ◊

For the sake of simplicity, from now on the scope of this text is restricted again to *unconditional* specifications only. However, all results can be transferred to conditional axioms in a similar way as it was sketched above.

4.4 Algebraic Programming

Under the notion of "Algebraic Programming", we summarize a growing collection of programming systems which try to integrate paradigms from term rewriting, functional programming and logic programming. Typical examples of such systems are SLOG ([Fribourg 85a]), BABEL ([Moreno, Rodríguez 88], or ALF ([Hanus 90]). A common feature of these systems is the use of *narrowing* as a mechanism for adapting the concept of a logical variable for functional programs. In order to achieve an effective algorithm, these languages restrict the syntactical form of the rewrite rules by a so-called *constructor discipline*.

In this section, we do not build up a direct relationship to one of the above-mentioned languages. Instead, it is shown how a restriction to constructor discipline can be combined with non-confluent rewriting. These so-called *constructor-based* specifications are of particular interest for this study, since they admit powerful mechanical checking of properties like DET-completeness and DET-additivity. In the framework of constructor-based specifications, an adaptation of the narrowing algorithm is studied, which forms the basis for algebraic programming techniques. The narrowing mechanism is also used in a later section for comparing nonconfluent rewriting with logic programming.

4.4.1 Constructor-Based Specifications

The so-called constructor-based specifications are of interest, because a large part of specifications used in practice fits into this class. A first remark in this direction was given in [Guttag 75], case studies with larger specifications also demonstrate this fact. Examples of such cases studies are [Geser 86], [Hussmann/ Rank 89], there are many others documented in the literature. The approach taken here allows us to omit a number of restrictions which are sometimes presupposed in the literature on algebraic programming: left-linearity, non-overlapping property, confluence. Termination of all rewriting sequences, however, is very useful (but not always necessary) for a successful algorithmic treatment. Even for specifications which are not DET-complete (constructor-complete, respectively), a sensible semantics can be given in this framework.

The starting point of the definition is the observation that there is a close relationship between DET-completeness and the notion of sufficient completeness at is was coined in [Guttag 75]. An even closer relationship exists between DET-completeness and the notion of constructor-completeness, as it was defined in [Huet, Hullot 82]:

> Let $C \subseteq F$ be a subset of the function symbols of $\Sigma = (S, F)$. Function symbols in C are called *constructors*. An equational specification $T = (\Sigma, E)$ is called *constructor-complete*, iff:
> $$\forall\ t \in W(\Sigma) \colon \exists\ t' \in W(\Sigma_C) \colon\ t =_E^* t'$$
> where $\Sigma_C =_{def} (S, C)$.

The notion of constructor-completeness can be easily adapted for nondeterministic specifications. For this purpose, the DET-axioms must be restricted in such a way that they designate a set of (deterministic) constructor operations. In the following, we assume that within $\Sigma = (S, F)$ a subset $C \subseteq F$ of constructors is designated. As a notation, constructors are marked by the keyword **cons** (instead of **func**) .

Note that nondeterministic constructors are excluded here. They are not necessary in general, since in multi-algebras some kind of "constructor" for nondeterministic sets of values always is available (by the set-building operations).

Within a constructor-based specification, it is not necessary to give explicit DET-axioms, if all constructors are understood implicitly as deterministic. For the inclusion rules, a particular syntactical shape is assumed (like in [Huet, Hullot 82]) which guarantees that the term algebra of constructor terms is free.

The syntactical restriction described in the next definition has been shown to be an acceptable compromise between an abstract description of a system and some kind of efficiency. Rather complex specifications can be written down within this restricted language. On the other hand, there exist experimental compilers (for the case of a confluent rule set), which generate relatively efficient code from such specifications, for instance [Geser, Hussmann, Mück 88], [Hanus 90].

Definition 4.13 (Constructor-Based Specification)

A specification $T = (\Sigma, R)$, $\Sigma = (S, F)$, where $C \subseteq F$ is the set of constructor operations, is called *constructor-based*, iff:

(1)   All axioms in R are of the form
$$f(t_1,\ldots,t_n) \to t$$
where $f \notin C$, $t_i \in W(\Sigma_C, X)$ for $1 \le i \le n$.

(2)   R does not contain DET-axioms. All models of T implicitly must fulfil the following axioms:
$$DET(c(x_1,\ldots,x_n))$$
for all constructors $c \in C$ (where $x_1, \ldots, x_n$ are pairwise disjoint variables).   ◊

Example 4.14

The specification DOUBLE' (example 2.3) can be written as a constructor-based specification:

spec C_DOUBLE
sort   Nat
cons   zero: → Nat,   succ: Nat → Nat
func   add: Nat x Nat → Nat,   double: Nat → Nat,
            zero_or_one: → Nat
axioms
            add(zero,x) → x,   add(succ(x),y) → succ(add(x,y)),
            double(x) → add(x,x),
            zero_or_one → zero,   zero_or_one → succ(zero)

end   ◊

In a constructor-based specification, we have:
$$t \in W(\Sigma_C, X) \iff T \vdash DET(t).$$
Therefore, a term can be tested for determinacy by a simple syntactical test.

From condition (1) it follows that the set of constructors is *free*, i.e.:

$\vdash DET(t1) \wedge \vdash t1 \rightarrow t2 \Rightarrow t1 = t2.$

As a consequence of this fact, the deduction rules (DET-D) and (DET-R) are no longer needed for derivations.

The property of DET-additivity is automatically given for constructor-based specifications:

Corollary 4.15

Every constructor-based specification is DET-additive (with respect to the implicit DET-axioms).

Proof: Consequence of theorem 2.11 and definition 4.13.   ◊

The test for DET-completeness is particularly simple for constructor-based specifications. Well-known methods for testing constructor-completeness can be adapted for this purpose.

Definition 4.16 (C-completeness)

A constructor-based specification $T = (\Sigma, R)$ $(\Sigma = (S, F))$ with constructors $C \subseteq F$ is called *C-complete*, iff:

$$\forall [f: s_1 \times \ldots \times s_n \rightarrow s] \in F\backslash C:$$
$$\forall t_1 \in W(\Sigma_C)_{s_1}, \ldots, t_n \in W(\Sigma_C)_{s_n}:$$
$$\exists \langle f(t_1',\ldots,t_n') \rangle \rightarrow \triangleright \in R, \sigma \in SUBST(\Sigma_C, X):$$
$$\sigma(f(t_1',\ldots,t_n')) = f(t_1,\ldots,t_n),$$

i.e. iff for any function symbol all potential arguments (seen as tuples of constructor terms) are covered by the argument pattern of some axiom.   ◊

Algorithms for a test of C-completeness have been described for instance in [Huet, Hullot 82], [Padawitz 83], [Kounalis 85].

In order to derive DET-completeness from C-completeness, an additional property is needed, which ensures that for any term at least one rewriting sequence terminates. The following specification, for instance:

spec   NT
sort   s
cons   a: → s, c: s → s
func   f: → s
axioms
       $f → c(f)$
end

is constructor-based and C-complete, but it is not DET-complete.

Definition 4.17 (Termination)

Let → be a reflexive and transitive relation on Σ-terms which forms a semi-congruence with respect to to the term-building operations (i.e. $t → t' \Rightarrow f(...,t,...) → f(...,t',...)$).

A term t is called →-*terminal*, iff there is no proper →-descendant of f, i.e.

$$\forall t' \in W(\Sigma): t → t' \Rightarrow t' = t.$$

The relation → is called *weakly terminating*, iff for every term t there is at least one terminal →-descendant of t, i.e.

$$\forall t \in W(\Sigma): \exists t' \in W(\Sigma): t → t' \wedge t' →\text{-terminal}.$$

The relation → is called *(strongly) terminating*, iff for all $t \in W(\Sigma)$:
There is no infinite sequence of terms $(t_i)_{i \in \mathbb{N}}$ where

$$t → t_0, \ t_i → t_{i+1} \text{ and } t_i \neq t \text{ with } i \in \mathbb{N}.$$

Strong termination implies weak termination.   ◊

For testing the strong termination of a term rewriting relation there exist a number of powerful criteria (see [Huet, Oppen 80], [Dershowitz 87] for an overview). Criteria for weak termination can be derived from these methods (by considering subsets of the rewrite rules). In general, no algorithm can exist, which decides the termination of an arbitrary term rewriting relation (even in the restricted case of constructor-based specifications).

Theorem 4.18

If a constructor-based specification T is C-complete and if the term rewriting relation $\rightarrow$ generated by T (with the calculus according to definition 2.4) is weakly terminating, then T is DET-complete.

Proof:

Because of the weak termination property for every ground term t there is a $\rightarrow$-terminal term t' such that $T \vdash t \rightarrow t'$.

For this term holds: $t' \in W(\Sigma_C)$. If otherwise there was a function symbol from $F\backslash C$ contained in t', then there would exist also a subterm of t' which has the form

$$f(t_1,\ldots,t_n)$$

with $t_i \in W(\Sigma_C)$ for $i \in \{1,\ldots,n\}$. Because of the C-completeness then an axiom could be applied to t', in contradiction to the $\rightarrow$-terminality of the term t'. ◊

For hierarchical constructor-based specifications, as they were defined in section 2.4, even the modularity properties can be easily checked by a syntactical condition: Every constructor should be specified within the specification where its target sort is introduced.

Theorem 4.19

Let $T = (\Sigma, R)$ be a hierarchical specification with constructor basis C. Let $T0 = (\Sigma 0, R0)$, $\Sigma 0 = (S0, F0)$ be the primitive part of T with constructor basis $C0 \subseteq F0$.
If
$$\forall\, [c: s_1 \times \ldots \times s_n \rightarrow s] \in C:\ s \in S0 \ \Rightarrow\ c \in C0\,,$$
then T is sufficiently complete and hierarchy-consistent.

Proof:

Let $t \in W(\Sigma_C)_s$, $s \in S0$. By induction on the term structure of t, the condition on the declaration of constructors yields $t \in W(\Sigma 0)$.
Let $t, t' \in W(\Sigma 0)$, $T \vdash t \rightarrow t'$. Because of definition 4.13 (1) no axiom out of $R\backslash R0$ can be applied to t. By induction on the derivation we have $T0 \vdash t \rightarrow t'$. ◊

It is quite obvious, how the calculus for rewriting with DET (definition 2.4) can be specialized to the case of constructor based specifications: The precondition of determinacy can be tested just by the syntactical criterion whether a term is built from the constructor symbols and variable symbols only.

Definition 4.20 (Constructor-Based Term Rewriting)

Given a specification $T = (\Sigma, R)$ with constructor basis C, a special case of the calculus from definition 2.4 is defined by:

(REFL), (TRANS), (CONG)   as in definition 2.4

(AXIOM-1-C)

$$\frac{}{\sigma l \rightarrow \sigma r} \qquad \text{if } l \rightarrow r \in R, \sigma \in SUBST(\Sigma_C, X).$$

Here Σ_C denotes the constructor-subsignature of Σ ($\Sigma_C = (S, C)$).
Derivations within this calculus are denoted by the symbol $\vdash_C$.   ◊

The following theorem establishes the connection between chapter 2 and the above-mentioned calculus.

Theorem 4.21

In a specification $T = (\Sigma, R)$ with constructor basis C, the following proposition holds for $t1, t2 \in W(\Sigma, X)$:
$$T \vdash t1 \rightarrow t2 \quad \Leftrightarrow \quad T \vdash_C t1 \rightarrow t2.$$

Proof:

The "$\Leftarrow$"-case (soundness) is a consequence of the fact that $\vdash_C$ is a special case of the calculus from definition 2.4 except of (AXIOM-1-C). Wherever (AXIOM-1-C) is applied, the condition: $\forall x \in X: \vdash DET(\sigma x)$ holds because of the implicit DET-axioms, therefore (AXIOM-1-C) can be replaced by an application of the original deduction rule (AXIOM-1).

"$\Rightarrow$"-case (completeness):
The proof is conducted by induction on the derivation. The following deduction rules can be excluded here: (DET-D) and (DET-R) (since no

rewrite rule can be applied to a pure constructor term, due to the form of the left hand sides). This means that deductions for formulas of the kind ⊢ DET(t) can use only (DET-X) and (AXIOM-2). For both these rules $t \in W(\Sigma_C, X)$ holds. Therefore all deductions for DET-formulas can be omitted; applications of (AXIOM-1) can be replaced by (AXIOM-1-C). The remaining deduction rules (REFL), (TRANS), and (CONG) are common to both calculi.   ◊

Please note that this establishes a soundness result for constructor-based rewriting, which holds independently of the C-completeness of the specification. In fact, a specification which does not possess the property of C-completeness can be given a reasonable semantics by regarding the missing cases as "undefined". This idea is followed in more depth below in chapter 6 on partiality in nondeterministic specifications. Chapter 6 below also contains a special section on constructor-based specifications (section 6.3). Let us state the main results from section 6.3 shortly in advance:

- There is a well-defined semantics for constructor-based specifications even without the condition of C-completeness ("partial constructor-based specifications").

- The appropriate deduction calculus for partial constructor-based specifications coincides with the calculus of constructor-based rewriting (definition 4.20). Soundness and weak completeness results hold.

- This leads to a sublanguage of nondeterministic specifications which does not need any checks for DET-completeness and DET-additivity (DET-completeness is avoided by partiality; DET-additivity is ensured by the constructor discipline).

For the detailed machinery behind these results, see chapter 6. The results have been reported already here, since they are useful for a comparison of constructor-based nondeterministic specifications with algebraic and logic programming.

To summarize this section: Constructor-discipline can be easily integrated into nondeterministic specifications. The resulting term rewriting calculus differs from classical term rewriting (by the restriction to constructor matchings in a rewrite step). However, this restriction is necessary to ensure soundness in the nondeterministic case; and it additionally covers an elegant treatment of partiality without any further modification of the calculus.

4.4.2. Narrowing without Confluence

The catch-word *"narrowing"* denotes an algorithm, which tries to solve a system of equations within a theory described by equational axioms. A standard assumption in this field of research is that the set of axioms form a confluent and terminating term rewriting system. The idea of narrowing goes back to [Slagle 74] and [Lankford 75], a first formulation of the algorithm is due to [Fay 79]. Like most literature on narrowing, the exposition given here is based on the description in [Hullot 80].

It is interesting that the narrowing relation (even for a confluent term rewriting system) may be non-confluent. This leads to the idea to use implementations of narrowing to get machine support for nonconfluent term rewriting. See chapter 5 on more details about this approach. Another important observation is that the correctness and completeness proofs for narrowing do not make any use of the confluence property of a term rewriting system. This means that narrowing can be carried over to non-confluent rewriting systems, at least for those cases, where the rewriting relation is sufficiently similar to the classical case. Below, we study narrowing in the framework of (partial) constructor-based nondeterministic specifications.

Narrowing adds to term rewriting systems an algorithm similar to Prolog's resolution method which computes an answer substitution for queries. Given a constructor-based term rewrite system, a *query* consists again of inclusion rules of the form:

$$t1 \rightarrow t2,$$

where free variables can occur in t1 and t2. (More complex queries consisting of a sequence of such rules are omitted here, they can be treated analoguously.) The algorithm now has to look for constructor-substitutions σ such that

$$R \models \sigma t1 \rightarrow \sigma t2.$$

Such a substitution is called a *solution*. A good algorithm should be able to enumerate all such solutions. In order to use the rewriting techniques developed above, the narrowing method tries to find instead a constructor-substitution σ such that

$$R \vdash_C \sigma t1 \rightarrow \sigma t2.$$

The connection to actual solutions then is given by soundness and completeness of constructor-based term rewriting. Since there is only a weak completeness result available for rewriting, we can only expect narrowing to be weakly complete (that is for t2 being a constructor term).

In fact, the narrowing process is very tightly coupled with rewriting. It uses a relation which can be deduced in a similar way to the rewriting relation. Basically, the matching process in the rewriting algorithm is replaced by syntactical unification; and out of the unification process a partial approximation to the answer substitution is computed and stored. Rewriting sequences can be "lifted" into narrowing sequences, without any regard to confluence or termination assumptions.

For describing the narrowing process, a new kind of formula is used:

Definition 4.22 (Narrowing Rule)

A *narrowing rule* is a triple consisting of two terms $t1, t2 \in W(\Sigma, X)$ of the same sort and a substitution $\sigma \in SUBST(\Sigma_C, X)$; it is denoted by
$$t1 \; \text{-}N\!\!\to_\sigma t2.$$
◊

Definition 4.23 (Narrowing)

A narrowing rule $t1 \; \text{-}N\!\!\to_\sigma t2$ is called deducible using a constructor-based rewrite system R (denoted by $R \vdash t1 \; \text{-}N\!\!\to_\sigma t2$) iff there is a deduction for $t1 \; \text{-}N\!\!\to_\sigma t2$ according to the following deduction system:

(REFL-N) $$\frac{}{t \; \text{-}N\!\!\to_\iota t} \qquad \text{if } t \in W(\Sigma, X)$$

(TRANS-N) $$\frac{t1 \; \text{-}N\!\!\to_\sigma t2, \; t2 \; \text{-}N\!\!\to_\tau t3}{t1 \; \text{-}N\!\!\to_{\tau\sigma} t3} \qquad \begin{array}{l} \text{if } t1, t2, t3 \in W(\Sigma, X), \\ \sigma, \tau \in SUBST(\Sigma_C, X) \end{array}$$

(CONG-N)
$$\frac{t_i \; \text{-}N\!\!\to_\sigma t_i{}'}{f(t_1, \ldots, t_{i-1}, t_i, t_{i+1}, \ldots, t_n) \; \text{-}N\!\!\to_\sigma f(\sigma t_1, \ldots, \sigma t_{i-1}, t_i{}', \sigma t_{i+1}, \ldots, \sigma t_n)}$$

if f $\in$ F with rank f: $s_1 \times \ldots \times s_n \rightarrow s$, $t_j \in W(\Sigma, X)_{sj}$,
$t_i' \in W(\Sigma, X)_{si}$, $\sigma \in SUBST(\Sigma_C, X)$

(AXIOM-N)   ———————

$$t \xrightarrow{N}_\sigma \sigma\rho r$$

if t $\in$ W(Σ, X), t $\in$ X, $\triangleleft \rightarrow \triangleright \in$ R,
ρ is a renaming such that $Vars(\rho) \cap Vars(t) = \emptyset$,
$\sigma \in SUBST(\Sigma_C, X)$ where σ is a *mgu* of t and ρl. $\Diamond$

This calculus formally defines the notion of narrowing and is consistent with the usual definitions. The only difference to the standard notion is that narrowing is restricted here to the generation of constructor-substitutions. Thus an implementation enumerating all derivations within this calculus can be gained from a classical narrowing implementation by a small modification. For constructor-based systems, the calculus describes only *innermost* narrowing steps. Moreover, for C-complete systems, the calculus coincides exactly with innermost narrowing, as it has been defined for instance in [Fribourg 85].

Example 4.24

Consider the following specification of sequences over an arbitrary data sort (we do not give any function symbols for this sort here), together with a "choice" operation:

spec SC
sort   Data, Seq
cons   empty: $\rightarrow$ Seq,   insert: Set $\times$ Data $\rightarrow$ Seq
func   choose: Seq $\rightarrow$ Data
axioms
        choose(insert(s,x)) $\rightarrow$ x,
        choose(insert(s,x)) $\rightarrow$ choose(s)
end

Please note that this is a partial constructor-based specification; it treats choose(empty) as undefined.

We have for instance the following narrowing derivations starting from the term choose(U)

(The variable U is a free variable to be considered as "unknown" in a query.)

|- choose(U) $-N\twoheadrightarrow_\sigma$ x   where σ (U) = insert(s,x)

|- choose(U) $-N\twoheadrightarrow_\tau$ y   where τ (U) = insert(insert(s,y),x).   ◊

Correctness of the narrowing method above means that all narrowing sequences are just "liftings" of rewrite sequences.

Lemma 4.25

Let T = (Σ, R) be a constructor-based specification, t1, t2$\in$W(Σ, X), $\sigma\in$SUBST(Σ_C, X) such that

   T |- t1 $-N\twoheadrightarrow_\sigma$ t2.

Then the following rewriting derivation exists:

   T |-$_C$ σt1 $\twoheadrightarrow$ t2.

Proof: By induction on the length of the derivation for |- t1 $-N\twoheadrightarrow_\sigma$ t2.   ◊

The following (rather technical) lemma shows that narrowing as defined above describes all "liftings" of a sequence of rewriting steps. This means, if there is a solution to a query (in the sense mentioned above), then it can be found with the narrowing method.

Lemma 4.26

Let T = (Σ, R) be a constructor-based specification, t1, t2$\in$W(Σ, X), V$\subseteq$X a set of "protected variables" with Vars[t1]$\subseteq$V, $\sigma\in$SUBST(Σ_C, X) such that Dom[σ]$\subseteq$V and

   T |-$_C$ σt1 $\twoheadrightarrow$ t2.

Then there are substitutions λ, $\sigma'\in$SUBST(Σ_C, X), a term t2'$\in$W(Σ, X) and a set of Variables V' with V$\subseteq$V'$\subseteq$X such that:

   T |- t1 $-N\twoheadrightarrow_{\sigma'}$ t2' and

(i)   Vars[t2']$\subseteq$V' $\wedge$ Dom[λ]$\subseteq$V' $\wedge$ Vars[σ']$\subseteq$V',

(ii) σ =$_{[V]}$ λ σ', and

(iii)   t2 = λ t2'.

Proof:

> The proof of this lemma (and the lemma itself) follows closely the
> ideas of [Hullot 80], which are described more extensively for instance
> in [Snyder 91].
>
> For the details of the proof see appendix A.   ◊

In order to get an algorithm for solving queries, liftings of rewriting sequences
are enumerated in such a way that a special case of every solution is reached. As
it was remarked above, the weak completeness result entails that this works only
well for queries t1 → t2, where q is a *constructor* term. The algorithm for
solving a query is then very similar to the classical narrowing algorithm: Its
main part is an enumeration of all possible narrowing sequences starting from
the left hand side of the query:

Algorithm 4.27 (Sketch)

> **Input:** R (a constructor-based rewrite system), ‹t1 → t2› (a query)
> **Output:** all possible solutions for the query
> **Method:**
> Search for all terms t2' and substitutions τ such that R |- t1 -N→$_\tau$ t2'
> holds.
> For all such terms and substitutions do:
> > If the term t2' is unifiable with the term τt2 (with mgu. μ),
> > then output $\mu\tau$ as a solution.   ◊

Note that a classical implementation of narrowing exactly performs the required
algorithm, if the query is reformulated as "t1 = t2". If t2 is a constructor term,
no narrowing steps can take place within it. So the only way to solve the
equation is by narrowing steps on t1 and by unification of the left and right hand
sides of the query.

Example 4.28

> Given the specification SC from example 4.24 and the query
> choose(U) → zero , the algorithm above will compute the following
> solutions:
> > [insert(s,zero) / U]
> > [insert(insert(s,zero),x) / U]
> > [insert(insert(insert(s,zero),x),y) / U]
> and many other solutions (in fact an infinite enumeration).   ◊

The correctness of the algorithm is an easy consequence of lemma 4.25, the completeness of the algorithm is formulated by the following theorem:

Theorem 4.29

Let $T = (\Sigma, R)$ be a constructor-based specification, $Q = [t1 \rightarrow t2]$ a given query where $t2 \in W(\Sigma_C, X)$.

If a substitution $\sigma \in SUBST(\Sigma_C, X)$ is a solution of Q, then there are substitutions $\lambda, \sigma' \in SUBST(\Sigma_C, X)$ such that σ' is computed by algorithm 4.27 and $\sigma =_{[V]} \lambda\sigma'$, where $V =_{def} Vars[t1] \cup Vars[t2]$.

Proof:

σ is a solution of Q ($\sigma \in SUBST(\Sigma_C, X)$)

$\Rightarrow$ $R \models \sigma t1 \rightarrow \sigma t2$

$\Rightarrow$ $R \vdash_C \sigma t1 \rightarrow \sigma t2$ (Thms. 4.21, 2.6)

$\Rightarrow$ $R \vdash t1 -N\rightarrow_\tau t2'$, $\sigma =_{[V]} \lambda'\tau$ and $\sigma t2 = \lambda' t2'$

                                                                                         (Lemma 4.26)

$\Rightarrow$ $R \vdash t1 -N\rightarrow_\tau t2'$ and $\lambda'\tau t2 = \lambda' t2'$ (Vars[t2]$\subseteq$V)

$\Rightarrow$ $R \vdash t1 -N\rightarrow_\tau t2'$ and $t2'$ is unifiable with $\tau t2$

                                                                                    (let μ be the mgu.)

$\Rightarrow$ The algorithm considers $t2'$ and τ and computes the solution
        $\sigma' = \mu\tau$ where $\lambda' = \lambda\mu$

$\Rightarrow$ $\sigma =_{[V]} \lambda'\tau = \lambda\mu\tau = \lambda\sigma'$.   ◊

To summarize, the concept of narrowing can be adapted for constructor-based nondeterministic specifications. Since the calculus of constructor-based term rewriting is sound and weakly complete for partial constructor-based specifications as well, also narrowing can be carried over to this special sublanguage.

The main advantages of partial constructor-based specifications are the presence of relatively powerful deduction techniques and the absence of any other than purely syntactical conditions - just the syntactic shape of constructor-based left hand sides of the axioms is sufficient. This interesting language has already been studied and used in a diffrent syntactical shape, within the framework of *logic programming*. The next section will show that we have reached now essentially a functional reformulation of classical logic programming.

4.5 Logic Programming

Logic Programming is a very successful paradigm of programming, in particular in applications of symbolic computation. The language Prolog is the most famous representative of logic programming. For our purposes, we will restrict ourselves to a small kernel of "pure Prolog" below.

The idea of logic programming is to use a purely logical framework for programming. This leads to an approach which can be located in between a true programming language and a specification language, sometimes also called "declarative programming". Logic programming to a large extent has been developed independently of the research in algebraic specification and term rewriting; for an overview of the theoretical background see [Lloyd 84]. However there exist very close relationships between these different worlds, which have been described for instance by [Deransart 83] and [Bosco et al. 88]. In this section, we will show a one-to-one correspondence between partial constructor-based nondeterministic specifications and classical definite logic programs. The results about narrowing in the nondeterministic framework above are needed here, as a functional equivalent to the concepts of logical variables and queries in logic programming.

The following definition summarizes a few of the most basic concepts of logic programming, which are needed for this section.

Definition 4.30 (Logic Program, SLD-Resolution)

> A (definite) *logic program* is built from terms over a signature Σ which contains two sorts, which are called here Data and Bool. There are only two kinds of operation symbols allowed: The *predicate symbols* and the *function symbols*, which are called *constructors* here. A predicate symbol p of arity n has the functionality
>
> $\quad$ p: Data $\times \ldots \times$ Data $\rightarrow$ Bool,
>
> a constructor c has the functionality
>
> $\quad$ c: Data $\times \ldots \times$ Data $\rightarrow$ Data.
>
> The terms of sort Bool ($W(\Sigma, X)_{Bool}$) are called *atoms*.
>
> A logic program consists of a finite set of *program clauses*, which are formulae of the shape
>
> $\quad$ H :-$\quad$ or $\quad$ H :- $B_1, \ldots, B_n$

where H, $B_1, ..., B_n$ are atoms. H is called the *head* of the program clause, the (possibly empty) sequence $B_1, ..., B_n$ is called its *body*.

A *query* is a nonempty sequence of atoms:

$$:- C_1, ..., C_n$$

where $C_1, ..., C_n$ are atoms.

A *goal* is a (possibly empty) sequence of atoms together with a substitution $\sigma \in SUBST(\Sigma, X)$, it is here written as

$$C_1, ..., C_n \text{ where } \sigma$$

The empty sequence of atoms in a goal is denoted as [].

Given a program, a goal can be transformed into another goal by the following (SLD-)*resolution* rule:

(RES) $A_1, ..., A_m, ..., A_k \text{ where } \sigma$

$$\theta A_1, ..., \theta A_{m-1}, \theta B_1, ..., \theta B_q, \theta A_{m+1}, ..., \theta A_k \text{ where } \theta\sigma$$

if $\langle A :- B_1, ..., B_q \rangle$ is a program clause,
θ is a *mgu* of A and A_m.

Remarks: This rule also can be used to replace an atom by an empty body, which shortens the goal. We did omit the technicalities of creating a variant of a program clause, which can be treated by applying a renaming (like in the narrowing calculus).

Given a query Q, a substitution σ is called an *answer*, iff, using this calculus

$$Q \text{ where } \iota \vdash [] \text{ where } \sigma$$

can be deduced.   ◊

4.5.1. Narrowing Simulates Logic Programming

As a first interesting correspondence, we will show that logic programming (in the simplistic sense of definition 4.30) can be simulated by narrowing. We use the framework developed above; however, the given simulation is independent of the extension to nondeterminism.

The idea is simply to encode a logic program as a set of rewrite rules working on the sort Bool. The "comma" operator is replaced by a logical "and".

Definition 4.31 (Translation of Logic Program)

Given a logic program P, a partial constructor-based specification $\Gamma(P)$ is associated to P, which is defined as follows.

The signature of $\Gamma(P)$ consists of the sorts
sort Data, Bool
and the function symbols
cons true, false: $\rightarrow$ Bool,
　　　　c: Data $\times \ldots \times$ Data $\rightarrow$ Bool　　　　for every constructor c in P,
func and: Bool $\times$ Bool $\rightarrow$ Bool,
　　　　p: Data $\times \ldots \times$ Data $\rightarrow$ Bool　　　　for every predicate p in P.

The axioms of $\Gamma(P)$ are the following:
　　　　and(true,true) $\rightarrow$ true,
　　　　and(false,x) $\rightarrow$ false,
　　　　and(x,false) $\rightarrow$ false,
　　　　$H \rightarrow$ and(B_1,and($\ldots$, B_n))
　　　　　　　　　　for every program clause $\langle H :- B_1, \ldots, B_n \rangle$,
　　　　$H \rightarrow$ true　　　for every program clause $\langle H:- \rangle$.

A query $Q = :- \langle C_1, \ldots, C_n \rangle$ is translated into
　　　　$\Gamma(Q) =_{def}$ and(C_1,and($\ldots$, C_n)).
This is extended to empty sequences of atoms by
　　　　$\Gamma([]) =_{def}$ true.　　　　　　　　　　　　　　　　$\Diamond$

Example 4.32

The following logic program is used to reverse lists. In difference to standard Prolog notation, we use the functions empty and cons to construct lists:
　　　　empty: $\rightarrow$ Data, cons: Data $\times$ Data $\rightarrow$ Data.
The program clauses are:
　　　　rev(L,R) :- rev1(L,empty,R)
　　　　rev1(empty,R,R)
　　　　rev1(cons(H,T),M,R) :- rev1(T,cons(H,M),R).
The corresponding specification is:

spec REV

sort   Data, Bool

cons   true, false: $\rightarrow$ Bool,
            empty: $\rightarrow$ Data, cons: Data $\times$ Data $\rightarrow$ Data.

func   and: Bool $\times$ Bool $\rightarrow$ Bool,
            rev: Data $\times$ Data $\rightarrow$ Bool,
            rev1: Data $\times$ Data $\times$ Data $\rightarrow$ Bool,

axioms
            and(true,true) $\rightarrow$ true,
            and(false,x) $\rightarrow$ false,
            and(x,false) $\rightarrow$ false,
            rev(L,R) $\rightarrow$ rev1(L,empty,R),
            rev1(empty,R,R) $\rightarrow$ true,
            rev1(cons(H,T),M,R) $\rightarrow$ rev1(T,cons(H,M),R)

end

Please note that this specification is truly partial: E.g. the term
            rev1(empty,empty,cons(1,2))
cannot be reduced to a Boolean constructor term. $\Diamond$

The following lemma makes the obvious relationship between resolution in P and narrowing in $\Gamma(P)$ explicit.

A purely technical remark: The lemma needs a more flexible use of the operator Γ, which transforms a sequence of atoms into a nested and-term. For the purposes of the proof, we consider such translations only *modulo* associativity of "and". Please note that this associativity is *not* added as an axiom to the specification $\Gamma(P)$, but is kept implicit within the proof. This is possible, since the structure of the and-term does not play any role in the narrowing computation; it is destroyed as soon as all the literals within it have been narrowed into "true".

Lemma 4.33

Let P be a definite logic program, $\Gamma(P)$ the associated specification. Then for any deduction by resolution from P holds:
If   Q **where** σ |- Q' **where** $\theta\sigma$
then $\Gamma(P)$ |- $\Gamma(Q)$ $\xrightarrow{N}_\theta$ $\Gamma(Q')$.

Proof:

For the case, where (RES) is applied, let $Q = \langle A_1, ..., A_m, ..., A_k \rangle$.

Then $\Gamma(Q) = \text{and}(A_1, \text{and}(..., \text{and}(A_m,...,A_k)))$.

According to (RES) there are a program clause $\langle A :- B_1, ..., B_q \rangle$ and a most general unifier θ of A and A_m.

In $\Gamma(Q)$ there is an axiom $\langle A \rightarrow \text{and}(B_1, \text{and}(..., B_q)) \rangle$. Using (AXIOM-N), we have

$$\Gamma(P) \vdash A_m \,\text{-}N\!\rightarrow_\theta \text{and}(\theta B_1, \text{and}(..., \theta B_q)).$$

Using (CONG-N) several times, we get

$$\Gamma(P) \vdash \text{and}(A_1, \text{and}(..., \text{and}(A_m,...,A_k))) \,\text{-}N\!\rightarrow_\theta$$
$$\text{and}(\theta A_1, \text{and}(..., \text{and}(\text{and}(\theta B_1, \text{and}(..., \theta B_q)),...,\theta A_k))),$$

which modulo associativity of "and" means

$$\Gamma(P) \vdash \Gamma(Q) \,\text{-}N\!\rightarrow_\theta \Gamma(Q').$$

There are two other cases to consider, which are implicitly contained in the resolution calculus (since it is described as a deduction system). They correspond to the reflexive and transitive closure. In fact this means a proof by induction on the length of the derivation.

The "reflexive" case (induction basis) is

Q where σ $\vdash$ **Q where** , i.e. $Q' = Q$, $\theta = \iota$.

Using (REFL-N), we get

$$\Gamma(P) \vdash \Gamma(Q) \,\text{-}N\!\rightarrow_\iota \Gamma(Q).$$

The "transitive" case (induction step) is

Q where σ $\vdash$ **Q'' where** $\theta'\sigma$ $\vdash$ **Q' where** $\theta''\theta'\sigma$,
$$\text{i.e. } \theta = \theta''\theta'.$$

By induction hypothesis, we have

$$\Gamma(P) \vdash \Gamma(Q) \,\text{-}N\!\rightarrow_{\theta'} \Gamma(Q''), \qquad \Gamma(P) \vdash \Gamma(Q'') \,\text{-}N\!\rightarrow_{\theta''} \Gamma(Q').$$

Using (TRANS-N), we get

$$\Gamma(P) \vdash \Gamma(Q) \,\text{-}N\!\rightarrow_{\theta'\theta''} \Gamma(Q'). \qquad\qquad \Diamond$$

Theorem 4.34

Given a definite logic program P and a query Q, any answer substitution σ, which is computed by SLD-resolution, is also a solution to the query $\Gamma(Q) \rightarrow \text{true}$ in the specification $\Gamma(P)$, and σ is computed by the narrowing algorithm 4.27.

Proof:

If σ is an answer to Q in the logic program P, there is a deduction

Q where ι |- [] where σ

Using lemma 4.34, then

$\Gamma(P)$ |- $\Gamma(Q)$ $\text{-}N\text{→}_\sigma$ $\Gamma([])$, and therefore

$\Gamma(P)$ |- $\Gamma(Q)$ $\text{-}N\text{→}_\sigma$ true.

This means that algorithm 4.27 will consider the subsitution σ, when working on the input $\langle\Gamma(Q) \to \text{true}\rangle$. Since true is trivially unifiable with true using ι, the substitution σ will be output as a solution.   ◊

Example 4.35

Consider the logic program from example 4.32. Using SLD-resolution, the query

:- rev(X,cons(1,cons(2,empty)))

is treated as follows (we show here only the relevant parts of the **where**-terms):

rev(X,cons(1,cons(2,empty))) **where** ι

|-   rev1(X,empty,cons(1,cons(2,empty))) **where** ι

|-   rev1(T1,cons(H1,empty),cons(1,cons(2,empty)))
                where [cons(H1,T1)/X]

|-   rev1(T2,cons(H2,cons(H1,empty)),cons(1,cons(2,empty)))
                where [cons(H1,cons(H2,T2))/X]

|-   [] **where** [cons(2,cons(1,T2))/X].

The corresponding narrowing sequences are:

rev(X,cons(1,cons(2,empty)))

$\text{-}N\text{→}_\iota$ rev1(X,empty,cons(1,cons(2,empty)))

$\text{-}N\text{→}[cons(H1,T1)/X]$

rev1(T1,cons(H1,empty),cons(1,cons(2,empty)))

$\text{-}N\text{→}[cons(H2,T2)/X]$

rev1(T2,cons(H2,cons(H1,empty)),cons(1,cons(2,empty)))

$\text{-}N\text{→}[2/H1,1/H2]$ true.   ◊

The theorem and the example show that SLD-resolution can be simulated by constructor-based narrowing in all operational details. Even the apparent difference that SLD-resolution has a more direct representation of the solution (as in the example above), comes only from different representation in the respectively calculi. Any implementation of narrowing will keep an analoguous "where-part", as it was shown above for SLD-resolution.

To summarize, SLD-resolution can be simulated by constructor-based narrowing. The translation has been proven sound for definite programs, but the possibility to specify also the result "false" for a predicate gives access to a simulation of the more general "normal programs" (in the sense of [Lloyd 84]). The next section shows that also SLD-resolution can be used to simulate constructor-based narrowing. Altogether this means that both mechanisms are essentially equivalent.

4.5.2. Logic Programming Simulates Narrowing

In this section another close correspondence between narrowing within constructor-based specifications and logic programming is studied. A technique is described which simulates the narrowing calculus from above using SLD-resolution. This can be used to construct a simple implementation of narrowing for nondeterministic specifications, in the partial constructor-based subcase.

The basic idea is here to translate every non-constructor operation into a predicate symbol. Nested occurrences of non-constructors are "flattened", using auxiliary variables, in order to achieve the syntactical form of definite Horn clauses. This technique has been studied in several variations. A good overview using a rather general approach can be found in [Bosco et al. 88]. The first usage of the technique was, according to this paper, in [Brand 74]. In the framework of logic programming with equality, the flattening technique has been studied in [Deransart 83], [Tamaki 84], [Barbuti et al. 85], [van Emden, Yukawa 87], and others.

Below the technique is sketched in a variant which is tailored to the particular subcase which is of interest here.

Definition 4.36 (Flattening)

Given a constructor-based specification $T = (\Sigma, R)$, $\Sigma = (S, F)$, where $C \subseteq F$ is the set of constructor operations, the *flattened signature* $\Phi(\Sigma)$ is defined as

$$\Phi(\Sigma) = (S \cup \{Bool\}, C \cup \{\Phi(f) \mid f \in F \backslash C\}),$$
where for $[f: s_1 \times \ldots s_n \rightarrow s] \in F$:
$$\Phi(f) = [f: s_1 \times \ldots s_n \times s \rightarrow Bool].$$

Terms from $W(\Phi(\Sigma), X)$ can be used to construct atoms of a logic program.

The *flattening* of a term gives a constructor term together with a sequence of atoms:

$$\Phi: W(\Sigma, X) \to W(\Sigma_C, X) \times (W(\Phi(\Sigma), X))^*,$$

$$\Phi[x] = (\, x, \varepsilon\,) \quad \text{if } x \in X,$$

$$\Phi[f(t_1,\ldots,t_n)] = (\, f(c_1,\ldots,c_n), B_1 \bullet \ldots \bullet B_n\,) \qquad \text{if } f \in C,$$

$$\Phi[f(t_1,\ldots,t_n)] = (\, z, f(c_1,\ldots,c_n,z) \bullet B_1 \bullet \ldots \bullet B_n\,) \qquad \text{if } f \in F \backslash C,$$

where $(\, c_i, B_i\,) = \Phi[t_i]$ $(i = 1,\ldots, n)$, $z \in X$ a "fresh" variable..

Using these operations, a logic program $\Phi(T)$ can be derived from the specification, if all sorts except of Bool are identified with Data, the predicate and constructor symbols are taken from $\Phi(\Sigma)$, and every axiom $\langle l \to r \rangle \in R$ is transformed into a program clause $\Phi[l \to r]$ according to:

$$\Phi[f(c_1,\ldots,c_n) \to r] = \langle f(c_1,\ldots,c_n,c) :\text{-} B \rangle,$$

$$\text{where } (\, c, B\,) = \Phi[r]. \qquad\qquad\qquad\qquad \Diamond$$

Example 4.37

The logic program $\Phi(C\_DOUBLE)$ associated to the constructor-based version of the "double"-specification (see example 4.14) is:

    add(zero,x,x) :-
    add(succ(x),y,succ(z_1)) :- add(x,y,z_1)
    double(x,z_2) :- add(x,x,z_2)
    zero_or_one(zero) :-
    zero_or_one(succ(zero)) :-.

The deduction of

    double(zero_or_one) $\to$ succ(succ(zero))

from example 2.5 can now be transformed into a resolution sequence within the logic program. We indicate the corresponding lines from example 2.7 in a separate column on the right side.

    :- double(z_3,X), zero_or_one(z_3)
    :- double(succ(zero),X)   (9)
    :- add(succ(zero),succ(zero),X)   (7)
    :- add(zero,succ(zero),z_3)   where [succ(z_3)/X]   (3)
    :- [] where [succ(succ(zero))/X]   (4)

However, there are also various other deductions admitted by the logic program. The standard strategy of a Prolog system would lead to:

:- double(z_4,X), zero_or_one(z_4)

:- add(z_4,z_4,X), zero_or_one(z_4)

:- add(z_5,succ(z_5),z_6), zero_or_one(succ(z_5)) where [succ(z_6)/X]

:- zero_or_one(succ(zero))   where [succ(succ(zero))/X]

:- [] $\Diamond$

It is an interesting observation that logic programming can expand the definition of double first, before expanding the definition of zero_or_one. This effect is due to the use of auxiliary variables, which lead to "structure sharing". For instance, the intermediate goal

:- add(z_4,z_4,X), zero_or_one(z_4)

cannot be represented within a term rewriting framework, since it involves a "sharing" of the result of zero_or_one. See section 5.2 for another aproach to structure sharing.

The soundness of the translation with respect to the intended semantics is almost obvious. To show the soundness, we adapt the interpretation of Horn clauses from logic programming (for instance :- is interpreted as reverse implication).

Theorem 4.38

Given a constructor-based specification $T = (\Sigma, R)$ and a model $A \in PMod(T)$[1], let the interpretation of a predicate symbol $\Phi(f)$ (for $f \in F$) be defined by

$$\Phi(f)^A (e_1, ..., e_n, e) \Leftrightarrow e \in f^A(e_1,..., e_n)$$

(where $e_1, ..., e_n, e$ are elements out of the respective carrier sets).

Then the axioms of $\Phi(T)$ are logically valid within A, i.e. $A \models \Phi(R)$.

Proof:

If the interpretation of a pair $(c, B_1 \bullet ... \bullet B_n)$, as it appears in the definition of Φ, is defined by

$$I_\beta^A[(c, B_1 \bullet ... \bullet B_n)] = \{e \in I_\beta^A [c] \mid I_\beta^A[B_1] \wedge ... \wedge I_\beta^A[B_n] \},$$

---

[1] PMod(T) denotes the class of models of T which admits partial and strict functions as interpretations of the functions. For a precise definition of PMod(T) see chapter 6 below.

then by induction on the structure of t, for an environment β the following can be shown:

$$I_\beta^A[\Phi[t]] = I_\beta^A[t].$$

For an inclusion rule $\langle f(c_1,\ldots,c_n) \to r \rangle$ in R, and $(c, B) = \Phi[r]$, this means

$$e \in I_\beta^A[r] \Leftrightarrow e \in I_\beta^A[\Phi[r]] \Leftrightarrow e \in I_\beta^A[c] \wedge I_\beta^A[B] \quad (*)$$

Since A is a model of T, we have

$$(\forall e: e \in I_\beta^A[r] \Rightarrow e \in I_\beta^A[f(c_1,\ldots,c_n)])$$

Therefore, using (*)

$$(\forall e: e \in I_\beta^A[c] \wedge I_\beta^A[B] \wedge e_i \in I_\beta^A[c_i] \Rightarrow e \in f^A(e_1,\ldots,e_n))$$

which implies (due to the definition of $\Phi(f)^A$)

$$(\forall e: I_\beta^A[B] \Rightarrow I_\beta^A[\Phi(f)(c_1,\ldots,c_n,c)]]).$$

This last line is exactly the semantical meaning of $\Phi[f(c_1,\ldots,c_n) \to r].\Diamond$

The completeness of the implementation of narrowing by flattened SLD-resolution can be shown directly in terms of the deductions (as the first derivation in the example above indicates). Detailed proofs for this can be found in the literature. The following lemma shows the idea for such a proof, adapted to the special case studied here.

Lemma 4.39

Under the preconditions of definition 4.36, let t1, t2∈W(Σ, X), θ∈SUBST(Σ, X). Let the flattenings of t1 and t2 be given by $\Phi[t1] = (c1, B1)$, $\Phi(t2) = (c2, B2)$.
Then the following implication connects derivations in T and $\Phi(T)$:

T ⊢ t1 $-N\to_\theta$ t2 $\Rightarrow$

          $\forall$ σ∈SUBST(Σ, X):   B1 **where** σ ⊢ B2 **where** θ'σ

where θ'∈SUBST(Σ, X) such that θ'c1 = c2, θ' = $\theta \cup \lambda$, and Dom[λ] contains only the variables introduced during the flattening of t1.

Proof: See appendix A.   ◊

Lemma 4.39 provides the main argument for the following theorem, which shows the close correspondence between constructor-based narrowing and SLD-resolution.

Theorem 4.40

Let $T = (\Sigma, R)$ be a constructor-based specification, $Q = [t1 \rightarrow t2]$ a given query where $t2 \in W(\Sigma_C, X)$.

Every solution $\sigma \in SUBST(\Sigma_C, X)$, which is computed by algorithm 4.27, is an answer of the logic program
$$\Phi(T) \cup \{ eq(X, X) :- \}$$
(where eq is a predicate symbol not used in $\Phi(T)$)
to the query
:- $eq(c1, t2) \bullet B1$ where $(c1, B1) = \Phi[t1]$.

Proof:

If σ is computed by the algorithm, there is a term $t2'$ such that $T \vdash t1 \rightarrow_N^\tau t2'$, $\sigma = \mu\sigma$ and μ is a mgu. of $\tau t2$ and $t2'$. Since $t2'$ must be a constructor term, $\Phi[t2'] = (t2', \varepsilon)$. Using lemma 4.39, in $\Phi(T)$ exists the resolution sequence B1 **where** $\sigma \vdash []$ **where** $\tau'\sigma$. This means for the query from above:
$eq(c1, t2) \bullet B1$ **where** $\iota \vdash eq(\tau'c1, \tau't2)$ **where** τ'
According to lemma 4.39, $\tau'c1 = t2'$ and $\tau't2 = \tau t2$. So we have
$eq(c1, t2) \bullet B1$ **where** ι
$\vdash \quad eq(t2', \tau t2)$ **where** τ'
$\vdash \quad []$ **where** $\mu\tau'$
(using (RES) on the program clause for eq).
This means that σ is an answer substitution.   ◊

As an example for an implementation of non-confluent rewriting on top of Prolog, see the LOG(F) system [Narain 88].

This completes the comparison between algebraic programming in nondeterministic specifications and logic programming. An almost one-to-one correspondence could be found in the subcase of partial constructor-based specifications.

From the logic programming viewpoint, this result can be understood as a way to subsume an important subcase of nondeterministic specifications. However, it should be kept in mind that the general case of nondeterministic specifications (and in particular partial ones, as studied below in chapter 6) provides a richer language than the simple sublanguage which can be translated to Prolog. For instance, it admits the definition of a deterministic basis which is different from a true subsignature, also reductions (equations) between terms of the deterministic basis are legal in the general case.

From the algebraic specification viewpoint, the result above leads to a simple implementation of deductions within nondeterministic specifications, for the above-mentioned sublanguage. The task of computing for a ground term t1 all ground constructor terms which fulfil $T \models t1 \rightarrow t2$, is accomplished by the query $[t1 \rightarrow X]$, where X is a "fresh" variable. However, in the Prolog implementation it is not easy to take advantage of the situation, where a subpart of a specification is given using a canonical system of rules. In term rewriting (and narrowing) implementations, it is easy to normalize intermediate terms and goals using a canonical subsystem (see lemma 5.10 below). As an alternative, the techniques described in [Cheong, Fribourg 91] are interesting, where "simplification" of intermediate goals in logic programming is studied.

Chapter 5

Implementation and Examples

For the practical use of a specification language, algorithmic support is essential. Software tools can be used for instance to test a given specification against informal requirements, to generate test data for an implementation, or to generate (semi-)automatically formal proofs for propositions over a specification. Below it is shown that existing tools for term rewriting can be used for experiments with nondeterministic specifications.

5.1 Term Rewriting

Most of the currently available interpreters for algebraic specifications provide an algorithm which reduces a term to normal form. In the following it is explained, how such algorithms can be generalized to non-confluent rewriting systems. It will turn out that there are basically two ways to do so: Using classical term rewriting with a particular strategy, or using graph rewriting techniques.

The question which has to be answered by a reduction algorithm is, for given ground terms t1 and t2, whether

$$T \models t1 \rightarrow t2.$$

As it was shown above in section 2.2, a complete deduction system for this task can be constructed only using semi-formal rules (for instance breadth-induction) or by a conditional calculus in the sense of [Walicki 92/92]. So we restrict our attention here to the case where a simple rewriting-like calculus has been shown to be sound and complete. This is the case for T |- DET(t2). Together with the weak completeness result, we are looking for ground terms t2 such that

$\qquad$ T |- t1 → t2 $\qquad$ where T |- DET(t2).

The calculus under discussion has been defined in definition 2.4 and is quite similar to term rewriting, except of one important point (which is necessary to achieve soundness in the nondeterministic case). The only difference between the classical term rewriting calculus and the calculus of definition 2.4 comes from the use of the DET-predicate and the corresponding deduction rules (DET-X, DET-D, DET-R, AXIOM-1, AXIOM-2). The deduction rules (DET-X), (DET-D), (DET-R) and (AXIOM-2) serve only for deducing DET-axioms, so the main difference is in (AXIOM-1), where a rewrite rule can be applied only if for all terms of the matching substitution determinacy has been proven (using the DET-predicate). The classical term rewriting mechanism must be modified in such a way that it respects this built-in restriction for substituting only determinate terms.

A first idea for avoiding the "built-in" predicate DET can be found in the analoguous situation for partial equational specifications (using a DEF-predicate). [Broy, Pair, Wirsing 84] propose to simulate the DEF-predicate by an operation with a Boolean result. Unfortunately, this technique *cannot* be transferred to inclusion rules and the DET-predicate. For instance, consider the following axioms:

$\qquad$ DET_OP(a) → true, DET_OP(b) → true, f → a, f → b.

In such a framework, with the rules (CONG) and (TRANS) the formula

$\qquad$ |- DET_OP(f) → true

can be deduced, which obviously is not always correct.

5.1.1 Innermost Rewriting

A better suited approach for a number of cases is the *innermost-strategy for replacement* which is well known from the operational semantics of applicative programming languages. Innermost-replacement means to apply a rewrite rule to a term t only, if no axiom can be applied to any subterm of t. [O'Donnell 77] explains that innermost-rewriting corresponds to a "call-by-value" semantics (cf. also [Bauer, Wössner 81]). Similarly, innermost-rewriting is appropriate for the "call-time-choice"-semantics, which is under consideration here. If the specification T is DET-complete, then for every ground term t the following inclusion holds:

No axiom can be applied to any subterm of t [1] $\Rightarrow$ T l- DET(t).

This means soundness of innermost rewriting with respect to definition 2.4. This idea is followed now in detail.

In the theory of term rewriting the following notions are known [O'Donnell 77]:

Definition 5.1 (Redex)

Let R be a term rewriting system.

An occurrence u$\in$Occ[t] within a term t is called a *redex*, iff there exists an axiom ⟨l → r⟩$\in$R and a substitution σ$\in$SUBST(Σ, X) such that t/u = σl. A redex u$\in$Occ[t] is called *innermost*, iff there is no further redex located in t below u, i.e. iff for all v$\in\mathbb{N}$*, v$\neq\varepsilon$:

u•v$\in$Occ[t] $\Rightarrow$ u•v is not a redex in t.

A term rewriting step t1 $\rightarrow_R$ t2 is called *innermost*, iff an axiom of R is applied at an innermost redex in t1. The restriction of the term rewriting relation to innermost rewriting is denoted by $\rightarrow_R^{im}$, its transitive closure by $\rightarrow_R^{im*}$, respectively. $\Diamond$

The relationship between innermost rewriting and nondeterministic rewriting over a deterministic basis can be made more precise (for ground terms) as follows.

Theorem 5.2

Let T = (Σ, R) be a DET-complete specification. Then for all t1, t2 $\in$ W(Σ):

$$t1 \rightarrow_R^{im*} t2 \quad \Rightarrow \quad T \text{ l- } t1 \rightarrow t2.$$

Proof:

As it was already mentioned, for every $\rightarrow$-terminal term t we have T l- DET(t). (Because of DET-completeness, there must be a t' such that T l- t → t' and l- DET(t'). Since t is terminal, the only possibility for this is the case using (REFL), where t = t'.)

---

[1] Such terms have been called *terminal* wrt. $\rightarrow$ above in definition 4.17.

In an innermost rewriting step using the matching substitution σ, the terms σx are terminal. Therefore T $\vdash$ DET(σx), which fulfils the additional condition of the deduction rule (AXIOM-1).   ◊

The following counterexample shows that the reverse direction does not hold:

Example 5.3

> **spec** FDT
> **sort**   s
> **func**   a: → s, g: → s,   f: s → s
> **axioms**
>   DET(a),   DET(g),
>   f(g) → a,   g → a
> **end**

The specification FDT is DET-complete and DET-additive. We have
>   FDT $\vdash$ f(g) → a,

but there is no innermost term rewriting sequence using the inclusion rules of FDT such that
$$f(g) \rightarrow_R^{im*} a. \qquad\qquad ◊$$

The example above shows that additional syntactical restrictions for the specifications are necessary to ensure not only soundness but also completeness of innermost term rewriting. A very simple but usable sublass of specifications is given by the *constructor-based specifications*, as defined in section 4.1.1. So the further argumentation in this section only refers to constructor-based specifications. For C-complete constructor-based specifications, the theorem above can be sharpened.

Theorem 5.4

Let T = (Σ, R) be a C-complete constructor-based specification. Then for all t1, t2 $\in$ W(Σ):
$$T \vdash t1 \rightarrow t2 \quad \Leftrightarrow \quad t1 \rightarrow_R^{im*} t2.$$

Proof:

Theorem 4.21 gives the result $T \vdash t1 \to t2 \Leftrightarrow T \vdash_C t1 \to t2$. So we can restrict our attention to constructor-based deductions (see definition 4.20 for $\vdash_C$).

The "$\Rightarrow$"-case is a consequence of the syntactic form of the axioms: All proper subterms of a $\vdash_C$-redex are automatically constructor terms; therefore no redices can be contained within them.

The "$\Leftarrow$"-case follows from theorem 5.2. $\Diamond$

These results can be reformulated immediately in the form of an algorithm. (We use here an informal notation, which should be self-explaining.)

Algorithm 5.5

Let $T = (\Sigma, R)$ be a C-complete specification with constructor basis C.

input: a ground term $t \in W(\Sigma)$
output: a ground term $t' \in W(\Sigma_C)$, such that: $T \vdash_C t \to t'$

funct reduce = (**term$_\Sigma$** t) **term$_\Sigma$**:
if $\exists\ u \in Occ[t1]$: $t/u = f(t_1,\ldots,t_n) \wedge f \notin C$
      then   **term$_\Sigma$** $t_1' = $ reduce(t_1); ...; **term$_\Sigma$** $t_n' = $ reduce(t_n);
                Choose nondeterministically an axiom $\langle l \to r \in R$
                and $\sigma \in SUBST(\Sigma_C)$, such that $\sigma l = f(t_1',\ldots,t_n')$;
                reduce($t[u \leftarrow \sigma r]$)
      else   t
fi $\Diamond$

Here the innermost reduction is realized by the control flow: All subterms of the given term are normalized (transformed into a constructor form) before the application of an axiom. The C-completeness guarantees that there always exists an appropriate axiom at occurrence u. Note that the algorithms works *nondeterministically with a non-determinate result*. It computes an *arbitrary* $t' \in W(\Sigma_C)$ such that the condition $\vdash_C t \to t'$ is fulfilled. For many cases, it will be a more realistic implementation to compute the *set* of all appropriate t'. (Then it is easy to determine whether a given t' occurs within this set.) The algorithm reduce is *correct* (i.e. all the possible results t' fulfil $\vdash_C t \to t'$) and *complete* (i.e. for each t' such that $\vdash t \to t'$ there exists a possible computation of reduce(t) with result t'). The correctness and completeness with respect to the semantical concepts of chapter 2 follows from the theorems 5.4, 2.6 and 2.15.

If considered as an algorithm for the computation of a result set, reduce is a *terminating* algorithm only, if the given axiom set generates a terminating term rewriting relation. For a non-terminating axiom set reduce can be considered as a *semi-algorithm*, which tries to generate the correct (infinite) result set. Such an enumeration obviously cannot terminate. (For implementation questions in this context see section 5.1.2 below.)

Given a terminating rewriting relation, reduce is a true (terminating) algorithm. In this case for any ground term t there exist only finitely many $t' \in W(\Sigma_C)$ such that $\vdash t \rightarrow t'$. As a consequence, then the $\rightarrow$-relation for arbitrary ground terms t1 and t2 is decidable, by a comparison of the (finite) sets of constructor normal forms:

Algorithm 5.6

> Let $T = (\Sigma, R)$ be a C-complete specification with constructor base C where the corresponding rewrite relation is terminating.
>
> input: Two ground terms $t1, t2 \in W(\Sigma)$
> output: Boolean value, indicating whether $T \vdash_C t1 \rightarrow t2$
>
> **funct** decide_$\rightarrow$ = (**term**$_\Sigma$ t1, **term**$_\Sigma$ t2)**bool**:
> $\quad\quad \forall$ **term**$_\Sigma$ t': t' $\in$ Results(reduce(t2))
> $\quad\quad\quad\quad\quad\quad\quad \Rightarrow$ t' $\in$ Results(reduce(t1)) $\Diamond$

The correctness and completeness of algorithm 5.6 relies on the theorems 2.19 and 5.4.

In the general case of possibly nonterminating rewriting sequences (see example 2.20), a breadth-induction proof can be necessary, in order to answer the question whether t1→t2 holds (even for ground terms t1, t2).

In the following two subsections, two essential issues will be adressed, in which an implementation for nonconfluent rewriting differs from usual implementation techniques for term rewriting. It turns out that the new problems are basically the same as they appear in an implementation of narrowing.

5.1.2 Search Strategies

The main difference between a classical interpreter for term rewriting systems and algorithm 5.5 comes from the fact that within a nondeterministic specification all admissible results are of interest, using arbitrary (nondeterministic) choices during the evaluation. For confluent rewriting, it is sufficient to study an *arbitrary* evaluation, since the final result here is independent of any actual choices during evaluation.

Example 5.7

The following reductions refer to the specification NAT from example 1.12 (which is made constructor-based by defining C={zero,succ}). We try to reduce the term add(some,some).

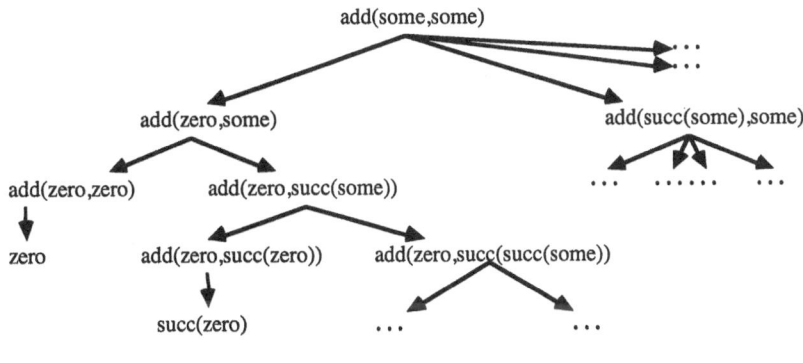

If a classical term rewriting interpreter (for confluent axiom sets) was applied to this example, it would deliver only one result. Whether this result is "zero" or "$succ^n(zero)$" (for an arbitrary n) or even nontermination (that is no result, actually), depends on the chosen evaluation strategy. An interpreter for nondeterministic specifications, however, should enumerate all the possible results (even if this is a nonterminating enumeration). Therefore the interpreter needs a tree-like organisation of rewriting sequences, as in the figure above; the corresponding tree search is similar to an interpreter for Prolog. In order to cope with nonterminating enumerations, a facility is needed to stop the interpreter after a finite number of results.

The example moreover demonstrates that an implementation of the tree search
by backtracking (*depth-first search*) like in Prolog is not always appropriate. If
the axioms are applied in a "strange" order, a depth-first search may follow an
infinite path, without ever reaching one of the results. For such cases a *breadth-*
first search seems to be more promising. A "breadth-first"-interpreter is
guaranteed to deliver every result after a finite amount of time (nevertheless it
may go into a nonterminating computation). For efficiency reasons, in many
cases a "depth-first"-organisation will be preferred, maybe with a preset limit on
the depth. This means that the user should have a choice between alternative
search strategies. Another interesting approach is to use an appropriate multi-
processor architecture for the evaluation of independent rewriting sequences in
parallel. (First attempts in this direction have been described for instance in
[Pinegger 87], similar ideas are followed in [Dershowitz, Lindenstrauss 90].)

It is clear that the nondeterministic rewriting algorithm is basically of
exponential complexity. Therefore those optimizations are particularly useful
which help to reduce the search space.

5.1.3 Optimizations

An essential step towards a smaller search space is achieved already by a special
innermost strategy. If the alternatives of one step are restricted to one singular
redex, for instance the leftmost one (*leftmost-innermost*), then the search space
from example 5.7 can be reduced to the following:

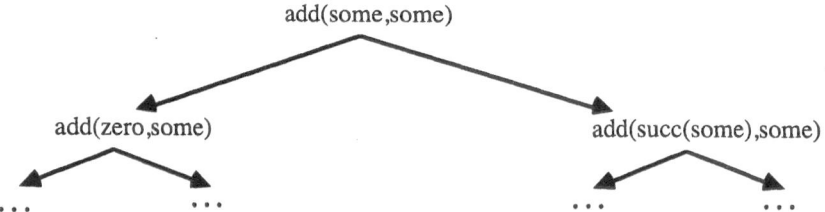

Please note that at every redex still the full range of applicable axioms is
considered. In this example, there exist two alternative axioms for every redex
(in comparison to 2*(number of redices) alternatives above). Again a similarity
to Prolog's SLD-resolution technique can be observed here: The *selection* of one
single clause from a goal is analoguous to the technique described here.

The optimization preserves correctness and completeness of the algorithm (since the replacements for various innermost redices can be considered independently).

Another important optimization refers to *confluence*. It should be ensured that an interpreter for nondeterministic specifications works well in the special case of confluent axioms. In this case it should achieve roughly the same efficiency as a classical interpreter. Unfortunately, the situation is more complicated, since the interpreter has to handle *mixed forms* of confluent and non-confluent rewriting. An extreme example is the following variant of the specification NAT:

Example 5.8

Let the axioms of a specification (intended for natural numbers) be:

$add(zero,x) \rightarrow x$, $\qquad\qquad\qquad$ $add(x,zero) \rightarrow x$,

$add(succ(x),y) \rightarrow succ(add(x,y))$, $\quad$ $add(x,succ(y)) \rightarrow succ(add(x,y))$,

$zero\_or\_one \rightarrow zero$, $\qquad\qquad\quad$ $zero\_or\_one \rightarrow succ(zero)$.

A tree of possible innermost rewritings then is given by:

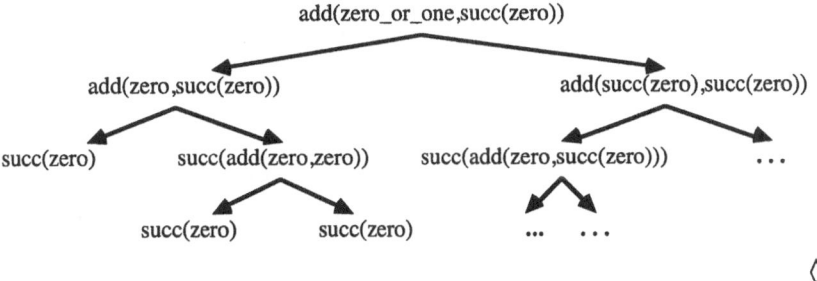

$\Diamond$

In this example, almost the full search effort goes into the inspection of superfluous paths, which are all equivalent to one out of two single paths. It should be sufficient to investigate these two paths. This idea goes into the direction of a remark in [O'Donnell 85] where an implementation of term rewriting is described (which does not yet support nondeterminism): "The ideal facility would allow equational definitions with multiple normal forms, but recognise special cases where uniqueness is guaranteed." (p. 135)

The obvious idea for the example above is to *normalize* the terms occuring in the tree with the subset of axioms which is terminating and confluent. In the example, the rules for the operation add can be used for normalization:

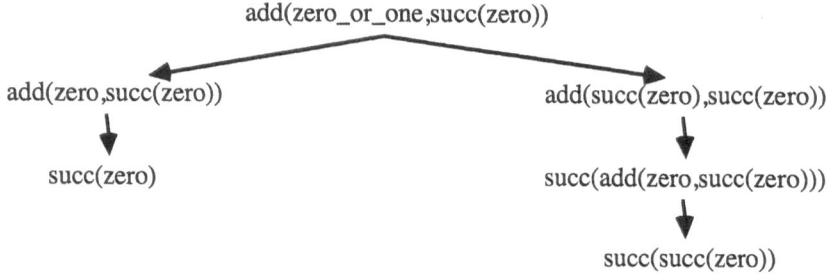

For the correctness of this optimization it is quite important that only those terms are normalized which contain deterministic operations only.

The correctness of such an optimization is obvious, as long as the principle of innermost rewriting is kept. For a proof of completeness, it has to be shown that the subset of axioms used for normalization cooperates with the rest of the axioms. More precisely: Let $D \subseteq R$ be a set of "deterministic" axioms, which shall be used for normalization, and therefore are terminating and confluent. Normalization induces an equivalence relation on ground terms:

$t \approx_D t'$ $\Leftrightarrow_{def}$ the D-normal forms of t and t' are equal.

Then we have to show the following "sub-commutativity property" :

$$t1 \to_{R \backslash D}^{im} t2 \;\wedge\; t1' \approx_D t1 \;\Rightarrow\; \exists\, t2': t1' \to_{R \backslash D}^{im} t2' \;\wedge\; t2 \approx_D t2' \,.$$

The lemma below shows that this property holds under appropriate preconditions. We use here the following notion of an innermost-normal form:

Definition 5.9 (Innermost Normal Form)

Let D be a term rewriting system such that $\to_D^{im*}$ is terminating and confluent.

The *innermost-normal form* $\downarrow_D^{im}[t]$ of a term $t \in W(\Sigma)$ is defined inductively by:

$$\downarrow_D^{im}[t] =_{def} \begin{cases} \downarrow_D^{im}[t[u \leftarrow \sigma r]] & \text{if } \exists u \in Occ[t], \triangleleft \to r \triangleright \in D, \\ & \sigma \in SUBST(\Sigma_C) : t/u = \sigma l \\ t & \text{otherwise.} \end{cases}$$

Lemma 5.10

Let $T = (\Sigma, R)$ be a constructor-based specification, $D \subseteq R$ a subset of the axioms such that the following conditions are satisfied:

- $\rightarrow_D^{im*}$ is terminating and confluent.

- D is *non-overlapping* with R\D, i.e. there does not exist any term t such that $\langle l_D \rightarrow r_D \rangle \in D$, $\langle l \rightarrow r \rangle \in R$, $\sigma \in SUBST(\Sigma_C)$ and $\sigma l_D = t = \sigma l$.

In this case, for t1, t2 $\in$ W(Σ):

$$t1 \rightarrow_{R\backslash D}^{im} t2 \;\Rightarrow$$

$$\exists\, t2': \downarrow_D^{im}[t1] \rightarrow_{R\backslash D}^{im} t2' \;\wedge\; \downarrow_D^{im}[t2] = \downarrow_D^{im}[t2'] .$$

Proof: See appendix A.   ◊

This lemma shows that the efficiency of confluent rewriting is not lost, when nonconfluent reduction is allowed. In order to designate a well-suited subset D, the following facts are to be observed:

- $\rightarrow_D^*$ is terminating $\Rightarrow$ $\rightarrow_D^{im*}$ is terminating

- $\rightarrow_D^*$ is confluent $\not\Rightarrow$ $\rightarrow_D^{im*}$ is confluent

Therefore, termination can be tested with one of the usual methods (see [Dershowitz 87]). The confluence of D-innermost-term rewriting does not follow directly from the confluence of D, but the classical method from [Knuth, Bendix 70] can be easily adapted to this case. (The critical pairs just have to computed with respect to innermost rewriting.)

Example 5.8 demonstrates that there is an important gain of efficiency by normalization if there are overlaps of the left hand sides of axioms within D. But even if this not is the case, and the number of terms is not reduced, the normalization optimization can still improve the efficieny. Since it is more expensive to represent a branching node ("choice point") of the proof tree on a real machine, memory space and time (for copying terms) can be saved.

5.2 Graph Rewriting

This section gives a discussion of a concept which is useful for the implementation of term rewriting in general, but in particular for non-confluent rewriting. The basic idea for this concept has been studied already in [Astesiano, Costa 79] for the semantics of nondeterministic processes ("Sharing in Nondeterminism"), but it can be generalized to arbitrary term rewriting systems. Term rewriting is performed here on terms which explicitly *share* some subterms. The term structure is enriched by information recording which pairs of equal subterms are *identical*. In [Hesselink 88], the combinator-like notation of terms ("accumulated arrows") leads to a similar effect.

From a completely independent aspect, term rewriting with sharing has been studied as an efficient implementation technique for (confluent) term rewriting. Starting from techniques for the implementation of the lambda-calculus (Wadsworth 1971), various approaches have been developed by Staples (1980), Raoult (1984), Barendregt et al. (1987), Hofmann and Plump (1988). In [Corbin, Bidoit 83], it is recommended to represent terms by *DAGs* (directed acyclic graphs), to achieve a simple and efficient implementation of unification and substitution algorithms on terms. In many implementations of term rewriting (among them RAP [Hussmann 85/87]) these ideas have been used successfully.

Unfortunately, the exact description of rewriting on graph-like structures leads to a significant technical overhead, if compared with term rewriting. Below, we reproduce some of the most important notions from [Barendregt et al. 87] and demonstrate the particularities of non-confluent rewriting in this context. In order to use the terminology of [Barendregt et al. 87] with only slight adaptations, the following arguments only apply to the case which is studied there. Therefore we assume here tha axioms to be left-linear and not to contain "extra variables" (which occur in the right hand side of a axiom, but not in the left hand side). The results can be generalized to remove these restrictions; but this generalization is not covered here.

5.2.1 Representation of Terms by Graphs

The following definition is almost literally taken from [Barendregt et al. 87].

Definition 5.11 (Graph)

Let $\Sigma = (S, F)$ be a signature, $X = (X_s)_{s \in S}$ a sorted set of variable names.

A *(labelled directed) graph (over Σ)* is a triple

$$G = (N, \text{lab}, \text{arg}),$$

where N is set of *nodes*, lab: $N \to F \cup X$ is the *labelling function*, and arg: $N \to N^*$ is the *argument* (or *successor*) function. The i-th component of arg(n) is denoted by $\text{arg}(n)_i$.

Th graph G is called *well-sorted*, iff there is a function sort: $N \to S$ such that for all $n \in N$:

$$\text{lab}(n) = x \wedge x \in X_s$$
$$\Rightarrow \text{sort}(n) = s,$$
$$\text{lab}(n) = f \wedge [f: s_1 \times \ldots \times s_k \to s] \in F$$
$$\Rightarrow \text{sort}(n) = s \wedge \text{sort}(\text{arg}(n)_i) = s_i$$

(for all $i \in \{1, \ldots, k\}$, i.e. $|\text{arg}(n)| = k$).

For two nodes n, $n' \in N$ the node n' is said to be *reachable from* n, iff either $n' = n$ or there is a $n'' \in N$, such that $\text{arg}(n)_i = n''$ and n' is reachable from n''.

The graph G is called *acyclic* iff every node is reachable from itself only through the trivial case in this definition.

A *rooted graph* is a quadruple $G' = (N, \text{lab}, \text{arg}, \text{root})$, where root $\in N$ and all nodes in N are reachable from root.

The *subgraph* of G at a node $n \in N$ is defined as the graph $G|n = (N_n, \text{lab}_n, \text{arg}_n)$ with node set $N_n = \{m \in N \mid m \text{ is reachable from } n\}$, $\text{lab}_n = \text{lab}|N_n$ and $\text{arg}_n = \text{arg}|N_n$.   ◊

For the purposes of term representation, we use rooted directed acyclic graphs (DAGs). It is obvious that a term is a subcase of such graphs. However, it is important to use only representations which *share variable occurrences*.

Definition 5.12 (Graph Representation of Terms)

Given a term $t \in W(\Sigma, X)$, a directed acyclic graph $GR[t] = (N, lab, arg, root)$ representing t can be constructed as follows (where $n_{<index>}$ is used as a name for a unique object out of some basic set of nodes).

$N = \{ n_u \mid u \in Occ[t] \wedge t/u \notin X \} \cup \{ n_x \mid x \in Vars[t] \}$;

lab: $N \rightarrow F \cup X$ such that

$\qquad x \in X \Rightarrow lab(n_x) = x$,

$\qquad t/u = f(t_1,\ldots,t_k) \Rightarrow lab(n_u) = f$;

arg: $N \rightarrow N^*$ such that

$\qquad x \in X \Rightarrow arg(n_x) = \varepsilon$,

$\qquad t/u = f(t_1,\ldots,t_k) \Rightarrow |arg(n_u)| = k \wedge arg(n_u)_i = n_{u \bullet i}$ for

$i \in \{1,\ldots,k\}$;

$\qquad root = n_\varepsilon \in N$.

Please note that for any variable x, there is only a single node in N labelled with x, which is called n_x. $\Diamond$

Example 5.13

The graphical notation for a graph is in most cases easier to understand and conceive than the formal notation from above.

The term add(x,x) is represented by the graph depicted at the right margin. Please note that the variable x is shared. The upper node is called n_ε in the formalism above, the lower one is called n_x.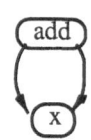

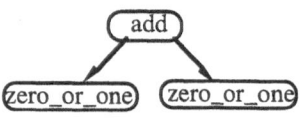

The graph at the right represents the term add(zero_or_one, zero_or_one). Please note that the arguments of add are not shared here. The upper node is called n_ε in the formalism above, the lower ones are called $n_{<1>}$ and $n_{<2>}$, respectively $\Diamond$

The following definition adds semantic interpretation to the notions of [Barendregt et al. 87]. The basic idea is here that every node of the graph is assigned to a single value (by a so-called valuation function). The nondeterministic breadth of interpretations is given by the range of such valuation functions for the graph. This is needed to achieve a sensible interpretation of graphs which share other subterms than just variables.

Definition 5.14

Given a model $A \in Mod(T)$ and an environment $\beta \in ENV(X, A)$, the interpretation $I_\beta^A[G]$ of a (rooted directed acyclic) graph $G = (N, lab, arg, root)$ is defined by

$$I_\beta^A[G] = \{ val(root) \mid val \in VAL_\beta^A[G] \}$$

where $VAL_\beta^A[G]$ denotes the set of admitted valuation functions. To be precise, it is a family of sets of functions, indexed by a sort. Again the sort is omitted for better readability.

$VAL_\beta^A[G] =$

$\{ val: N \to s^A \mid$
$\forall n \in N: (lab(n) = x \wedge x \in X \Rightarrow val(n) = \beta x) \wedge$
$\qquad (lab(n) = f \wedge f: s_1 \times \ldots \times s_k \to s] \in F \Rightarrow$
$\qquad\qquad val(n) \in f^A(val(arg(n)_1), \ldots, val(arg(n)_k)) \}$.

The well-definedness follows from the fact that G is acyclic.

It is obvious that $I_\beta^A[GR[t]] = I_\beta^A[t]$.   ◊

5.2.2 Rewriting of Term Graphs

The definition of a graph replacement rule again is mainly taken from [Barendregt et al. 87] (with correction of a minor error). Additionally, a formal translation of inclusion (or term rewrite) rules into graph rewrite rules is given.

Definition 5.15

A *graph rewrite rule (over Σ)* is a triple $(G, root_L, root_R)$ where G is a graph and $root_L$ and $root_R$ are nodes of G such that every node of G is reachable from either $root_L$ or $root_R$.

Given an inclusion rule $⟨l \to r⟩$ over Σ and X, a graph rewrite rule $GR[⟨l \to r⟩] = (G, root_L, root_R)$ is defined by:

$G = (N, lab, arg)$,

$N = \{ l_u \mid u \in Occ[l] \wedge l/u \notin X \} \cup \{ r_u \mid u \in Occ[l] \wedge r/u \notin X \}$
$\qquad \cup \{n_x \mid x \in Vars[l] \cup Vars[r] \}$;

$lab: N \rightarrow F \cup X$ such that

$\qquad x \in X \Rightarrow lab(n_x) = x$,

$\qquad l/u = f(t_1,...,t_k) \Rightarrow lab(l_u) = f$,

$\qquad r/u = f(t_1,...,t_k) \Rightarrow lab(r_u) = f$;

$arg: N \rightarrow N^*$ in analogy to definition 5.12;

$root_L = l_\varepsilon \in N$; $root_R = r_\varepsilon \in N$.

The application of a graph rewrite rule $(G, root_L, root_R)$ to some target graph $G_0 = (N_0, lab_0, arg_0, root_0)$ is defined as follows.

A *redex* for the rule in G_0 is a graph homomorphism ϕ: $G\lvert root_L \rightarrow G_0$, i.e. a function ϕ: $N \rightarrow N_0$, such that for all $n \in N$, which are reachable from $root_L$ holds: $lab(n) \notin X \Rightarrow lab_0(\phi(n)) = lab(n)$ and $arg_0(\phi(n)) = \phi^*(arg(n))$, where ϕ^* is the elementwise extension of ϕ to sequences of nodes.

Given such a redex, the application of the rule proceeds in three phases:

(i) *build phase:* We assume that nodes and variables of G and G_0 are disjoint. Then the right hand side of G is added to G_0, instantiating variables according to ϕ. This gives a new graph $G_1 = (N_1, lab_1, arg_1, root_1)$, formally:

$N_1 =$

$\qquad N_0 \cup \{n \in N \mid n$ reachable from $root_R$ and not from $root_L \}$,

$lab_1(m) = \begin{cases} lab_0(m) & \text{if } m \in N_0, \\ lab(m) & \text{otherwise,} \end{cases} \qquad$ for $n \in N_1$;

$arg_1(m)_i = \begin{cases} arg_0(m)_i & \text{if } m \in N_0, \\ arg(m)_i & \text{if } m, arg(m)_i \in N \cap N_1, \\ \phi(arg(m)_i) & \text{if } m \in N \cap N_1, arg(m)_i \notin N \cap N_1, \end{cases}$

$root_1 = root_0$.

The root for the instantiated right hand side is now $n_r \in N_1$, where

$\qquad n_r = \begin{cases} \phi(root_R) & \text{if } root_R \text{ reachable from } root_L, \\ root_R & \text{otherwise.} \end{cases}$

(ii) *redirection phase:* All references to $\phi(root_L)$ are replaced by references to n_r. This gives a new graph $G_2 = (N_2, lab_2, arg_2)$, formally:

$N_2 = N_1$,

$lab_2(m) = lab_1(m)$,

$$arg_2(m)_i = \begin{cases} n_r & \text{if } arg_1(m)_i = \phi(root_L), \\ arg_1(m)_i & \text{otherwise}, \end{cases}$$

$$root_2 = \begin{cases} n_r & \text{if } root_1 = \phi(root_L), \\ root_1 & \text{otherwise}. \end{cases}$$

(iii) *garbage collection phase:* Nodes which are not accessible from $root_2$ are removed. Formally this gives a graph $G_3 = G_2|root_2$.

A graph rewriting step is denoted by $G_0 \to_{GR} G_3$.   ◊

Example 5.16

In this example, the rule

add(zero, x) → x

is applied to the graph G_0, which contains a shared occurrence of the function symbol zero.

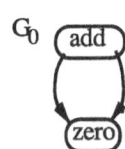

The rule is represented by the graph G. This is a special case, since the right hand side does not contain any non-variable nodes.

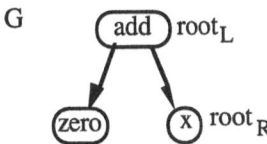

The result of the pattern matching of G onto G_0 is represented by the graph homomorphism ϕ:

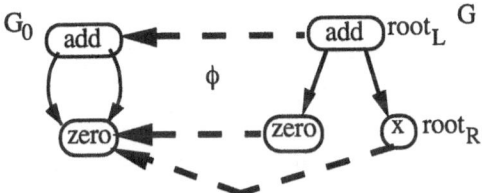

The graph G_1 does not differ much from the original graph G_0, since there is no node added. However, the nodes $\phi(root_L)$ and n_r indicate the nodes to be raplaced.

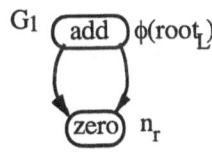

After the replacement, n_r becomes the root of
the graph G_2 (which does not differ from G_1 in
other respects). After garbage collection, G_3
contains one single node.

G_3 (zero)

Using the algebra NN from example 1.15, where the interpretation of
the function symbols is: $\text{zero}^{NN} = \{0, 1\}$, $\text{add}^{NN}(e1, e2) = \{e1+e2\}$,
we have the following interpretations for the graphs:
$$I^{NN}[G_0] = I^{NN}[G_1] = \{0, 2\},$$
$$I^{NN}[G_2] = I^{NN}[G_3] = \{0, 1\}.$$ ◊

5.2.3 Soundness and Completeness

Example 5.16 shows clearly that the standard graph rewriting techniques are *not
sound* in the general case for nondeterministic interpretations. This is due to the
non-injective mapping φ, which does not ensure that all the rule nodes which are
mapped onto a single target node, are always interpreted equally.

Again, the restriction to constructor-based specifications helps to overcome the
problem. In this case all the nodes mapped by φ are interpreted deterministically
(except of the root), and therefore graph rewriting is sound for this subcase. The
following theorem formulates soundness of graph rewriting in this sense.

Theorem 5.17

Let $T = (\Sigma, R)$ be a constructor-based specification, where all axioms
are left-linear and where in each axiom the variables in the right hand
side form a subset of the variables in the left hand side.
Let G_0 be a rooted acyclic graph over Σ. Let G_3 be a graph constructed
out of G_0 using the graph representation $GR[l \rightarrow r]$ of a rule $l \rightarrow
r \in R$, according to definition 5.15.
Given a model A and an environment β, we have
$$\forall \, val_3 \in VAL_\beta^A[G_3] \colon \exists \, val_0 \in VAL_\beta^A[G_0] \colon val_0(root_0) = val_3(root_3).$$

Proof:

The proof is technically rather complex. In appendix A, a sketch is
given which covers the essential arguments.   ◊

Please remember that the restrictions which were put onto the form of the axioms are only due to the use of the framework of [Barendregt et al 87]; they can be removed by extending this framework.

In [Barendregt et al. 87], it is shown moreover that also the completeness of graph rewriting with respect to term rewriting is not obvious. In general, there are term rewriting sequences which cannot be simulated by graph rewriting. This is illustrated by the following example.

Example 5.18 (Example 5.4 from [Barendregt et al. 87])

> **spec** CGR
> **sort** s
> **func** a: → s, b: → s, c: → s,
> f: s × s → s, g: s → s
> **axioms**
> f(a,b) → c, a → b, b → a,
> g(x) → f(x,x)
> **end**

> There is a term rewriting sequence
> $$g(a) \to f(a,a) \to f(a,b) \to c;$$
> but there is no graph rewriting sequence starting from any graph representation of g(a) and leading to a graph representation of c. The reason is that the first rule (f(a,b) → c) can never be applied, due to the sharing of the subterms instantiated for x. ◊

This counterexample has close similarities to the running example of this text (using the "double" operation). In fact, we can construct a model which shows that classical term rewriting is unsound for this example, under nondeterministic interpretations. This means that it is no longer a counterexample in the framework of this text.

Example 5.19

> A nondeterministic model C for the specification CGR from example 5.18 is given by:
> $s^C = \{\, a1, a2, c \,\}$,
> $a^C = b^C = \{\, a1, a2 \,\}$, $c^C = \{\, c \,\}$,

$$f^C(a1, a1) = \{ a1 \}, \qquad f^C(a1, a2) = \{ c \}, \qquad f^C(a1, c) = \{ c \},$$
$$f^C(a2, a1) = \{ c \}, \qquad f^C(a2, a2) = \{ a2 \}, \qquad f^C(a2, c) = \{ c \},$$
$$f^C(c, a1) = \{ c \}, \qquad f^C(c, a2) = \{ c \}, \qquad f^C(c, c) = \{ c \},$$
$$g^C(a1) = \{ a1 \}, \; g^C(a2) = \{ a2 \}, \; g^C(c) = \{ c \}.$$

Within the model C, the inclusion ‹g(a) → c› does not hold:
$$I^C[g(a)] = \{ a1, a2 \}, I^C[c] = \{ c \}. \qquad\qquad \Diamond$$

This demonstrates that the counterexample cannot be carried over to the nondeterministic case. Even better, under the preconditions of the above soundness result, also completeness holds with respect to constructor-based term rewriting. In order to state this result formally, the notion of "unravelling" ([Barendregt et al. 87]) a graph into a term is needed.

Definition 5.20

Let $G = (N, lab, arg, root)$ be an acyclic and finite graph over a signature Σ and variable names X.

The operation TM constructs a term $TM[G] \in W(\Sigma, X)$ out of G, according to the following definition:

$$lab(root) \in X \qquad \Rightarrow \qquad TM[G] = x,$$
$$\text{where } lab(root) = x;$$
$$lab(root) \notin X \qquad \Rightarrow \qquad TM[G] = f(TM[G_1],...,TM[G_n]),$$
$$\text{where } lab(root) = f, [f: s_1 \times ... \times s_n \to s] \in F,$$
$$G_i = G|arg(root)_i \text{ for } i \in \{1,...,n\}. \qquad \Diamond$$

Theorem 5.21

Let $T = (\Sigma, R)$ be a constructor-based specification, where all axioms are left-linear and where in each axiom the variables in the right hand side form a subset of the variables in the left hand side.

Then for any two terms $t1, t2 \in W(\Sigma, X)$ holds:
$$T \vdash_C t1 \to t2 \qquad \Rightarrow$$
$$\exists \text{ graph } G2: \; GR[t1] \xrightarrow{*}_{GR} G2 \wedge TM[G2] = t2,$$

where graph replacement refers to the rules $\{GR[\triangleleft \to \triangleright] \mid \triangleleft \to \triangleright \in R\}$.

Proof:

The proof mainly relies on the fact that the graph representation GR[t1] and subsequent rewriting steps always produce graphs in which all shared subgraphs are irreducible (with respect to graph rewriting). This is formalized by the following two predicates.

Let $G = (N, lab, arg)$, $n \in N$. Then

is_shared[n] $\Leftrightarrow_{def}$

$\exists\ n1, n2 \in N, i, j \in \mathbb{N}: n1 \neq n2 \wedge arg(n1)_i = n = arg(n2)_j,$

wf[G] $\Leftrightarrow_{def}$

$\forall\ n, n' \in N:$ is_shared[n] $\wedge$ n' reachable from n $\Rightarrow$

$lab(n') \in C \cup X.$

Using the predicate wf, the following lemma can be shown for an arbitrary acyclic and finite graph G1 over Σ and X:

$$wf[G1] \wedge TM[G1] = t1 \wedge l\text{-}_C t1 \to t2 \Rightarrow$$

$$\exists\ G2: wf[G2] \wedge TM[G2] = t2 \wedge G1 \xrightarrow{*}_{GR} G2.$$

The theorem then follows from the simple facts that wf[GR[t1]] and TM[GR[t1]] = t1.

In the appendix A, a sketch for the proof of the lemma is given. $\Diamond$

To summarize, we have shown that an implementation by graph rewriting is sound for constructor-based specifications, and that it is able to reproduce all the derivations which are admitted in constructor-based rewriting. It should be mentioned that graph rewriting does even admit a greater number of sound derivations than constructor-based rewriting or the innermost strategy for classical term rewriting (see section 5.1). Since graph rewriting is sound for constructor-based nondeterministic specifications (theorem 5.17), an arbitrary redex selection strategy can be used. An outermost replacement sequence corresponding to example 1.12 is, for instance, the following one. Please note that this reduction sequence is *not* deducible using constructor-based rewriting.

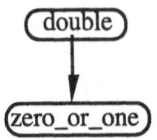

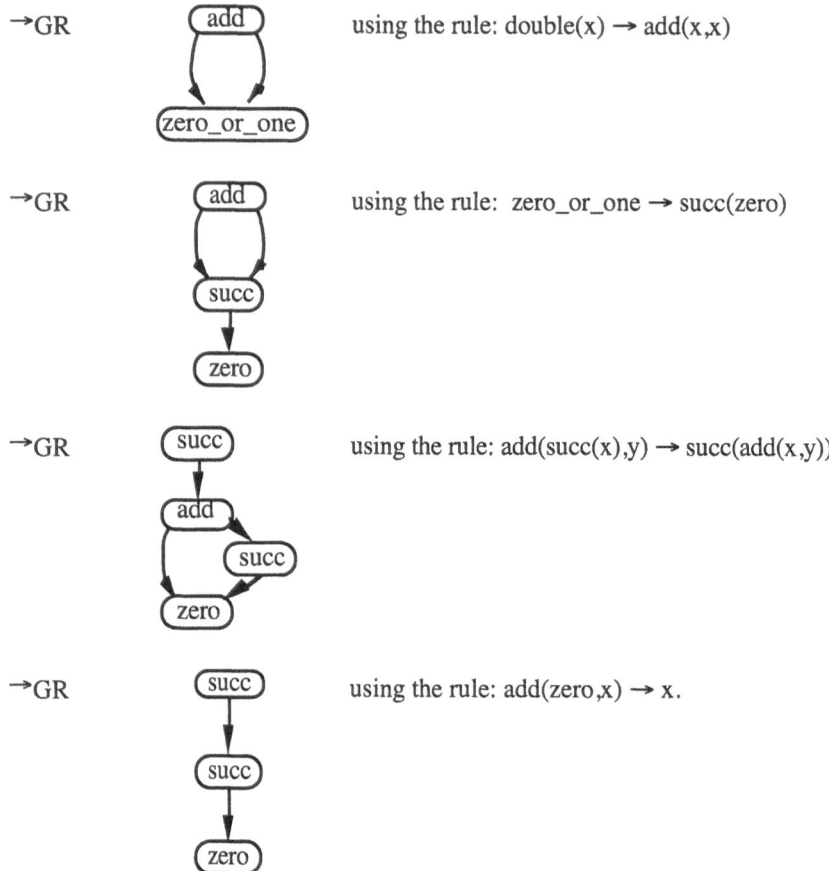

→GR   using the rule: double(x) → add(x,x)

→GR   using the rule: zero_or_one → succ(zero)

→GR   using the rule: add(succ(x),y) → succ(add(x,y))

→GR   using the rule: add(zero,x) → x.

An implementation of terms by directed acyclic graphs admits another optimization. If a subterm is changed (for instance by normalization), these changes may concern simultaneously many copies of the subterm. This behaviour is similar to the D-evaluation rule invented by Vuillemin (see [Bauer, Wössner 81]). Efficient implementations of graph reduction techniques are described for instance in [Johnsson 84].

It is also interesting to compare the implementation by graph rewriting with an implementation based on logic programming, as it was used for instance for example 4.37 above or in the LOG(F) system ([Narain 88]). Using the technique of translation to logic programs, a form of subterm sharing is present "for free", by Prolog's built-in variable sharing. This is the reason, why the logic programming and graph rewriting approach both show more flexibility in the reduction strategy than constructor-based rewriting.

To summarize, existing software tools can be used to perform deductions within nondeterministic algebraic specifications, if

- only constructor-based specifications are studied, and
- the implementation admits either
  - innermost term rewriting and a constructor-completeness test, or
  - representation of terms by graphs with variable-sharing.

The system RAP [Hussmann 85/87], although designed independently of the nondeterministic framework, fulfils the requirements from above (since it uses a graph representation for terms), so it can be used for computer experiments based on nondeterministic specifications.

5.3 Examples

This section shows the application of nondeterministic specifications to a few typical examples. The examples are taken from different areas of computer science; in order to keep the length of the examples within a reasonable size, only the basic ideas are sketched here. The first two examples are from theoretical computer science, then two classical examples for nondeterministic algorithms are given, and finally it is sketched how nondeterminism can be used to specify abstractly some concrete sequences of events within an operating system.

5.3.1 Nondeterministic Finite State Automata

Automata theory frequently uses nondeterministic machine models. The following example shows that this classical nondeterministic framework can be specified easily by algebraic methods.

In the following, a nondeterministic finite automaton is considered, which appears during the systematic construction of an algorithm for string pattern matching (cf. also [Knuth, Morris, Pratt 77]). The idea is here, to follow nondeterministically ("simultaneously") all possible patterns during the read process. The following simple automaton comes out of the task of checking whether one of the patterns ‹OL› or ‹LO› appears within a sequence of binary digits ({O,L}*):

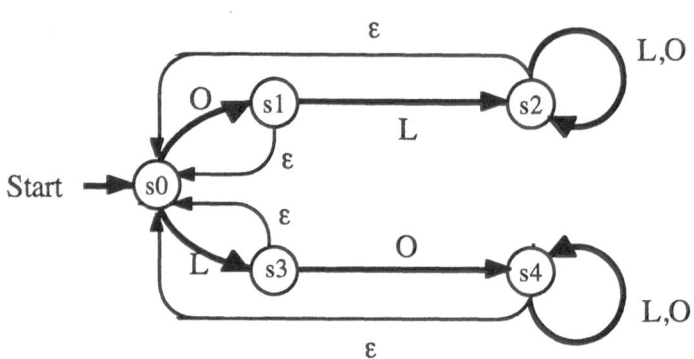

The symbol ε here denotes a so-called "spontaneous" transition. Please note that the automaton is constructed easily from the pattern matching task: For every pattern, a sequence of states and transitions is built, and from every state a spontaneous return into the start state is added. States s3 and s4 are "final states": Once one of these states has been entered by the automaton, the state cannot be changed furthermore, regardless of the input.

The corresponding (constructor-based) specification contains sorts for the states and for the input symbols. The automaton is coded into axioms for the one-step transition relation. This relation is described by the operation trans, which computes a follow state out of a state and an input symbol. trans* models the transitive closure of this transition relation. For the description of trans*, sequences of input symbols have to be considered, too.

spec NFA
sort   Input, State, Seq
cons   O, L: → Input,
            empty: → Seq,   append: Seq × Input → Seq,
            s0, s1, s2, s3, s4: → State
func   trans: State × Input → State,
            trans*: State × Seq → State

axioms

$trans(s0,O) \rightarrow s1,$ $trans(s0,L) \rightarrow s2,$

$trans(s1,L) \rightarrow s3,$

$trans(s2,O) \rightarrow s4,$

$trans(s3,O) \rightarrow s3,$ $trans(s3,L) \rightarrow s3,$

$trans(s4,O) \rightarrow s4,$ $trans(s4,L) \rightarrow s4,$

$trans(s1,x) \rightarrow trans(s0,x),$ { ε-transitions }

$trans(s2,x) \rightarrow trans(s0,x),$

$trans(s3,x) \rightarrow trans(s0,x),$

$trans(s4,x) \rightarrow trans(s0,x),$

$trans^*(s,empty) \rightarrow s,$

$trans^*(s,append(t,x)) \rightarrow trans^*(trans(s,x),t)$

end

The first block of axioms describes the transitions of the automaton which process an input symbol, the second one gives the spontaneous transitions. The third block of axioms (which is not specific for this particular automaton) serves for the derivation of the transitive closure of the transition relation.

The C-completeness of trans* is obvious, for trans the C-completeness can be seen from the fact that the following formulae are deducible:

$trans(s0,O) \rightarrow s1$ $trans(s0,L) \rightarrow s2$

$trans(s1,O) \rightarrow s1$ $trans(s1,L) \rightarrow s3$

$trans(s2,O) \rightarrow s4$ $trans(s2,L) \rightarrow s2$

$trans(s3,O) \rightarrow s3$ $trans(s3,L) \rightarrow s3$

$trans(s4,O) \rightarrow s4$ $trans(s4,L) \rightarrow s4$

The axioms of NFA are terminating, since there is no way to derive a cyclic sequence of ε–steps.

As an example for computations within NFA, consider the term

$trans^*(s0,append(append(append(append(empty,L),O),L))$

which specifies the set of automaton states after processing the input ‹LOL› (with start state s0). In a computer experiment, RAP computes the following set of states as a set of possible simplifications of this term:

{ s2 , s3 , s4 }

This can be interpreted as follows: After having processed the input, the automaton is in one of the states s2, s3 or s4. The appearance of s3 and s4 shows that the (overlapping) occurrences of both patterns (‹OL› and ‹LO›) in the input have been recognised.

This example clearly illustrates the purpose of a nondeterministic specification language: It does not help for the development of an efficient implementation (for this purpose, the automaton should be transformed into a deterministic one by the well-known methods), but it allows us to describe an inherently nondeterministic problem in a rather abstract and problem-oriented way.

5.3.2 Petri Nets

In order to demonstrate that also classical models of nondeterministic and distributed computing can be modelled using nondeterministic algebraic specifications, we give here a method for encoding Petri nets. In chapter 7, a more complex study of distributed computing (aiming at the language CSP) can be found. Here, we use a rather simplistic way of encoding Petri nets by a sort describing explicitly the state of the net. Other (mainly equivalent) ways of description from the literature ([Kaplan 88], [Meseguer 92]) also could be transferred to this framework.

We use the following Petri net taken from [Kaplan 88] to demonstrate the encoding. The Boolean Petri net shows a variant of the famous "producer-consumer" problem.

One possible idea of the encoding is to define a sort with a single constructor (net) keeping the information which of the places of the net is occupied by a token. Each transition is translated into an axiom which transforms such net states according to the firing rules for Boolean Petri nets. Each input place for the transition must contain a token, each output place must be free. The result state has tokens in all output places, and the tokens from the input places are removed.

spec PN
sort Token, Net
cons Y, N: Token,
       net: Token × Token × Token × Token × Token → Net
             { i-th argument of net corresponds to place i in the net }
func trans: Net → Net,   trans*: Net → Net
axioms
       trans(net(N, Y, p3, p4, p5)) → net(Y, N, p3, p4, p5),   {prod}
       trans(net(Y, N, N, p4, p5)) → net(N, Y, Y, p4, p5),   {send}
       trans(net(p1, p2, Y, Y, N)) → net(p1, p2, N, N, Y),   {rec}
       trans(net(p1, p2, p3, N, Y)) → net(p1, p2, p3, Y, N),   {cons}
       trans*(n) → n,   trans*(n) → trans*(trans(n))
end

Every state of the net reachable from the start configuration shown in the picture can be computed by reducing the term
       trans*(net(N, Y, N, Y, N))

An example is:
       trans*(net(N, Y, N, Y, N))
→   trans*(net(Y, N, N, Y, N))   {prod}
→   trans*(net(N, Y, Y, Y, N))   {send}
→   trans*(net(N, Y, N, N, Y))   {rec}
→   trans*(net(N, Y, N, Y, N))   {cons}
→   net(N, Y, N, Y, N).

Please note that the specification contains two sources of nondeterminism. The first one is the function trans*, which nondeterministically computes every reachable state from a given configuration. The other one is the function trans itself, which is specified in a non-confluent way:
       net(Y, N, Y, Y, N) ← trans(net(N, Y, Y, Y, N) → net(N, Y, N, N, Y)

Generally, nondeterministic specifications allow us to express concurrency directly, avoiding the technicalities of *interleaving* sequences, as it was observed in [Meseguer 92].

5.3.3 The Eight Queens Problem

It is a classical programming problem to place 8 queens on a checker board such that all of them are safe, such that no queen can be attacked by another one. The problem has been mentioned in [Manna 70] as a typical example for a nondeterministic program. Nondeterminism in this problem means, to find *some* position of the pieces which fulfils the safety condition.

The specification of the Eight Queens problem gives an example for the modular construction of nondeterministic specifications by hierachies as well as for the use of conditional rules. Basically, the specification follows the ideas proposed in [Manna 70].

Let BOOL and INT be given specifications for the truth values and for the integer numbers. These specifications are deterministic (in the sense that in every maximally deterministic model all operations have to be deterministic). Based on INT, finite sequences of integer numbers can be described by

spec SEQ_INT
basedon INT
sort   SeqInt
cons   empty: $\rightarrow$ SeqInt,   app: SeqInt $\times$ Int $\rightarrow$ SeqInt
func   length: $\rightarrow$ SeqInt
axioms
             length(empty) $\rightarrow$ zero,   length(app(s,x)) $\rightarrow$ succ(length(s))
end

A configuration of the chess pieces on a n$\times$n-board are described as a sequence of n natural numbers (the specification uses integer numbers), where the i-th number denotes, in which row the queen for column i is placed. The specification TEST defines the condition which is necessary for an extension of a correct configuration on a n$\times$n-board to a correct configuration on the (n+1)$\times$(n+1)-board: The queen placed in column n+1 must not be on a row or a diagonal which is already "occupied" by another queen. The specification uses the fact that a diagonal is characterized by a fixed value for the sum or difference of the row and column indices:

spec TEST
basedon INT, BOOL, SEQ_INT

func   ok: SeqInt × Nat → Bool ,

        samerow, samediag1, samediag2: SeqInt × Int → Bool

axioms

        samerow(empty,r) → false,

        samerow(app(s,r1),r2) → or(equal(r1,r2),samerow(s,r2)),

        samediag1(empty,d) → false,

        samediag1(app(s,r),d)

                → or(equal(add(r,succ(length(s))),d),samediag1(s,d)),

        samediag2(empty,d) → false,

        samediag2(app(s,r),d)

                → or(equal(sub(r,succ(length(s))),d),samediag2(s,d)),

        ok(s,r) → and(not(samerow(s,r)),

                    and(not(samediag1(s,add(r,succ(length(s))))),

                        not(samediag2(s,sub(r,succ(length(s))))))))

end

The Boolean term ok(s,r) is true iff r is a safe position for the n+1-th queen, where the configuration of the first n queens is given by s.

The nondeterministic specification itself now is rather simple: We look for a sequence of length 8 which has been constructed step by step according to the criterion above and which contains only numbers between 1 and 8:

spec QUEENS

basedon TEST, SEQ_INT, INT, BOOL

func   queens: →SeqInt,   qu: SeqInt → SeqInt

axioms

        queens → qu(empty),

        less_equal(length(q),8) → false ⇒ qu(q) → q,

        less_equal(length(q),8) → true & less_equal(0,r) → true &

        less_equal(r,8) → true & ok(q,r) → true

                ⇒ qu(q) → qu(app(q,r))

end

Please note that this specification already takes a rather operational viewpoint by describing the incremental construction of a solution (which in fact leads to a pruning of the search space). An even more optimized specification (or better program), tuned towards lazy nondeterministic rewriting, can be found in [Narain 88].

In the initial model of QUEENS, the interpretation of the term queens exactly contains a representation of the solutions for the Eight Queens problem. Experiments with the RAP system show that this specification is executable, but that optimizations (like for instance in [Narain 88]) are urgently needed. On a SUN SPARCstation 10 workstation with 32 MByte RAM the first solution for the problem in the formulation from above is found after approx. 140 CPU-seconds. An analoguous five queens problem needs 2 CPU-seconds for the first solution, all solutions are found within approx. 55 seconds.

5.3.4 The Monkey-Banana Problem

The next example also comes from [Manna 70], inspired by McCarthy. It is again a search problem, but, according to its origin from the field of "artificial intelligence", it is formulated as an experiment in animal behaviour.

The experiment is as follows: A monkey sits within a room, where a high box is placed on the floor and where a banana is fixed at the ceiling in a height unreachable for the monkey. The room contains nothing else. The following possibilities now are available for the monkey in order to get the banana: It can move on the floor of the room, it can climb on the box, and it can move the box around on the floor. The solution is (obviously) to move the box under the banana and then to climb onto the box.

A specification of this problem mostly contains the definition of trivial data structures. The specification WORLD contains the atomic objects which are needed for the problem description (vertical and horizontal positions, actions):

spec WORLD
sort VPos, HPos, Action
cons floor, ceiling: VPos,
        monkey_pos, banana_pos, box_pos: HPos,
        walk, carry: HPos → Action,   climb: Vpos → Action
end

The constants monkey_pos, banana_pos, box_pos mean the initial positions of monkey, banana and box on the floor. (Other positions are of no interest for this problem.)

The following specification describes states for the description of situations:

spec STATE
sort State, Pair
cons state: Hpos × VPos × Hpos → State,   pair: Action × State → Pair
end

The states are to be interpreted as follows: A triple state(mh,mv,b) describes the actual position of the monkey (mh in horizontal, mv in vertical dimension), as well as the (horizontal) position of the box (b). Pairs of actions and states are used to represent the relationship between a particular action and its consequences. The specification OPERATIONS lists which actions are admitted in a given state and which is the subsequent state after the action. Nondeterminism is used here to describe the choice between alternative actions:

spec OPERATIONS
basedon WORLD, STATE
func do: State → Pair
axioms
     do(state(x,floor,z)) → pair(walk(hp),state(hp,floor,z)),
     do(state(x,floor,x)) → pair(carry(hp),state(hp,floor,hp)),
     do(state(x,y,x)) → pair(climb(vp),state(x,vp,x))
end

These three rules describe exactly how the "world" can be changed by actions of the monkey. The monkey can move horizontally to an arbitrary position hp; in this case it remains on the floor (first rule). It can move the box on the floor to an arbitrary position, provided it is on the floor and at the same horizontal position as the box (second rule). It can climb onto the box (or down from it), if it is at the same horizontal position as the box (third rule).

The last specification part describes how sequences of actions are performed (and recorded) and how the goal of the game is defined. Let a specification SEQ_ACTION for sequences of actions be given, analoguously to 5.3.3.

spec STEPS
basedon WORLD, STATE, SEQ_ACTION, OPERATIONS
func   steps: State → SeqAction
a x i o m s
        do(s) → pair(a,s1) ⇒ steps(s) → append(a,steps(s1)),
        steps(state(banana_pos,ceiling,z)) → empty
end

The search itself means an enumeration of all reductions of the term
        steps(state(monkey_pos,floor,box_pos)) .

The RAP system finds the (optimal) constructor term
        append(walk(box_pos),
        append(carry(banana_pos),
        append(climb(ceiling),empty)))
after 0.08 CPU seconds (on the machine configuration described above, and using all possible optimizations).

5.3.5 Printer Scheduling

The last example is intended to give an impression, how nondeterministic specifications can be used for an abstract description of phenomena out of more practical fields of computer science. Below, a small sub-aspect of an operating system is described: the distribution of a sequence of printing jobs under a given number of printers (*scheduling*). For the sake of abstractness, any reference to a notion of time is avoided.

In this example, a printer always has one out of the two possible states: free or busy.

spec PRINTER_STATE
sort   PrinterState
cons   busy, free: → PrinterState
end

A printer is called busy if it is currently working on a printing job. At an arbitrary, unpredictable time the printing job is finished. At this moment the state of the printer changes from "busy" to "free". To capture this behaviour, below an operation next_status is specified nondeterministically, which is used to ask for the current state of a printer. The operation gets the current (old) state of the printer as its argument. It has two possibilities for its result: Either it delivers the argument state (printer state unchanged) , or it spontaneously declares the printer to be free (but only if the old state was "busy"). In this case, a printing job has been finished since the last query for the printer state.

spec PRINTER
basedon PRINTER_STATE
func   next_status: PrinterState $\rightarrow$ PrinterState
a x i o m s
        next_status(s) $\rightarrow$ s,
        next_status(s) $\rightarrow$ free
end

Two rather trivial specifications describe sequences of printing jobs (represented by natural numbers) and sequences of events during the execution of printing jobs. Events in this sense are: "Job i starts on printer j" and "Job i has to wait for a free printer".

spec JOB_QUEUE
basedon NAT
sort   JobQueue
cons   empty_Job: $\rightarrow$ JobQueue,
        append_Job: Nat × JobQueue $\rightarrow$ JobQueue
end

spec EXEC_QUEUE
basedon NAT
sort   ExecQueue
cons   empty_Exec: $\rightarrow$ ExecQueue,
        append_Exec: Nat × Nat × ExecQueue $\rightarrow$ ExecQueue,
        wait: Nat × ExecQueue $\rightarrow$ ExecQueue
end

The actual administration of the printers now is described by a specification which is deterministic except of its use of the nondeterministic operation next_status. The function scheduler transforms a sequence of jobs into a sequence of events. For this purpose, it uses an auxiliary operation sched, which gets the actual states of two printers as its arguments (fixing here the number of printers to the value 2). A job can be executed only if a free printer is ready for it, otherwise it has to wait:

spec SCHEDULER

basedon NAT, JOB_QUEUE, EXEC_QUEUE,
         PRINTER, PRINTER_STATE
func   scheduler: JobQueue → ExecQueue,
          sched: JobQueue × PrinterState × PrinterState → ExecQueue
axioms
          scheduler(q) → sched(q,free,free),
          sched(empty_Job,s1,s2) → empty_Exec,
          next_status(s1) → free ⇒
            sched(append_Job(jn,q),s1,s2) → append_Exec(jn,1,sched(q,busy,s2)),
          next_status(s2) → free ⇒
            sched(append_Job(jn,q),s1,s2) → append_Exec(jn,2,sched(q,s1,busy)),
          next_status(s1) → busy & next_status(s2) → busy ⇒
            sched(append_Job(jn,q),s1,s2)
                   → wait(jn,sched(append_Job(jn,q),s1,s2))
end

The following term describes the possible sequences of events for a sequence of three jobs:

          scheduler(append_Job(1,append_Job(2,append_Job(3,empty_Job))))

If RAP is called to reduce this term, it starts to enumerate a nonterminating list of event sequences; among them we find the following ones:

  append_Exec(1,1,append_Exec(2,1,append_Exec(3,1,empty_Exec)))
  append_Exec(1,1,append_Exec(2,2,append_Exec(3,1,empty_Exec)))
  append_Exec(1,1,append_Exec(2,2,wait(3,append_Exec(3,1,empty_Exec))))
  append_Exec(1,1,append_Exec(2,2,wait(wait(append_Exec(3,2,empty_Exec)))))

The first sequence can be understood as a sequence of three "very short" jobs which are finished before the next job arrives. In the second case, two jobs are given to two printers for parallel processing; the first job finishes first, and before the third job arrives. In the last two cases the third job has to wait for a free printer. It is interesting that an intuitive understanding of the results is easier when a time-oriented formulation is chosen. Nevertheless, the formal specification completely abstracts from the notion of time.

This example also can be used to demonstrate the use of narrowing for nondeterministic specifications. For instance, the following equation (with unknown variables j, j1, e1) can be understood as the question whether a sequence of events is admissible which starts with a waiting state:

scheduler(j) = wait(j1,e1)

RAP correctly does not find any solution (and terminates rapidly). The next equation asks whether it is possible that another job is printed before the first one in the job queue is started:

scheduler(append_Job(1,j)) = append_Exec(j1,p1,append_Exec(1,p2,e1))

Two solutions are found:

j = append_Job(1,*0), j1 = 1, p1 = 1 and

j = append_Job(1,*0), j1 = 1, p1 = 2

(where *0 is a system-generated variable, which stands for an arbitrary natural number). An interpretation of this solution is: The equation can be fulfilled only, if the job number of the first job is equal to the number of the second one. Here it can be seen, how an experiment with a software tool uncovers problems or mistakes within a specification: The specification above does not contain an axiom to exclude explicitly the multiple use of the same job number within a job queue.

Chapter 6

Partial Nondeterministic Specifications

Up to this point, only nondeterministic specifications have been considered the models of which are total algebras. The axioms were restricted to (conditional) inclusion rules. An important advanced concept for classical equational specifications is an appropriate treatment of "undefined" situations during the computation of a value. In order to show that partial nondeterministic specifications do not evoke essential new problems, a generalization of our approach to partial specifications is addressed here. The treatment of partiality follows tightly the concepts in [Broy, Wirsing 82].

6.1 Partial Operations

Partiality is a basic phenomenon in programming. Typical examples are algorithms which cannot terminate for all inputs (like an interpreter for a universal programming language) or overflow and underflow situations (for instance in arithmetic).

6.1.1 Undefined "Values"

A basic idea for the treatment of partial operations is that there are no "undefined values" in the sense of actual values, but that undefinedness means the *non-existence of a value*. Intuitively, such a situation is imagined best as a nonterminating computation. If this point of view is taken, the actual carrier set of a model can contain only defined (existent) values.

In general, the theory of partial operations presupposes all operations to be *strict*. Intuitively spoken, strictness describes the property that the non-existence of some argument value prohibits the computation of any result value. There exist generalizations of the theory to the case of *non-strict* operations (which obey some *monotonicity* restrictions, cf. [Möller 82], [Broy 87]). Intuitively, non-strict operations can compute a result in some cases from an incomplete set of argument values without "waiting" for some unnecessary argument. The work presented here does not cover such generalizations. But it can be assumed that even non-strict operations can be integrated well into the nondeterministic approach.

In analogy to section 1.1.1, a number of alternative approaches are discussed first, how to model partial nondeterministic operations mathematically.

Let f: s1 × s2 → s be a function symbol, s1, s2 and s sort symbols of some underlying signature.

Within an algebra A, let $s1^A$, $s2^A$ and s^A denote the respective carrier sets. The symbol ⊥ is used within the description of the operations to denote undefined situations. This "pseudo value" ⊥ is *not* a member of the carrier sets! For an arbitrary set M we use the abbreviation:
$$M^\perp =_{def} M \cup \{\perp\}$$

Using the terminology of [Broy, Wirsing 81], there are a number of alternatives for the interactions between ⊥ and defined values:

(a) Erratic Nondeterminism

> Let A be a model of the given specification,
> $$f^A: s1^A \times s2^A \to \wp^+((s^A)^\perp),$$
> where f^A is continued *strictly*, i.e.:
> $$f^A(\perp,e2) = \{\perp\} \text{ and } f^A(e1,\perp) = \{\perp\}$$
> (where $e1 \in s1^A$, $e2 \in s2^A$).

Here ⊥ is treated similarly to the defined values. In some situations there can be a choice between ⊥ and a set of defined values ("choice nondeterminism").

[Nipkow 86] and [Hesselink 88] use this form of partial operations. The other two approaches described below do not allow such a choice between "undefined" und "defined":

(b) Demonic Nondeterminism

Let B be a model of the given specification,
$$f^B: s1^B \times s2^B \rightarrow (\wp^+(s^B))^\perp,$$
where f^B is continued strictly, as above.

An informal explanation of the "demonic" approach is that an operation attempts to compute the whole set of *all possible results*. If the computation of one of these results does not terminate, the whole set of results is undefined. This can be observed from the fact that the definition above does not admit a choice (during the computation) between defined results and "undefined". Demonic nondeterminism is often called also "backtracking" nondeterminism, since it appears naturally in search procedures which are described nondeterministically.

From a model A, according to approach (a), a model B, according to approach (b) can be constructed by defining (for a term t):

$$I^B[t] = \begin{cases} \{\perp\} & \text{if} \perp \in I^A[t] \\ I^A[t] & \text{otherwise} \end{cases}$$

(c) Angelic Nondeterminism

Let C be a model of the given specification,
$$f^C: s1^C \times s2^C \rightarrow \wp(s^C),$$
where f^C is continued strictly, and where $\emptyset$ is considered as the representation for "undefinedness".

The approach of angelic nondeterminism does not handle undefined values explicitly. It tries to avoid undefinedness wherever possible, instead. Only if none of the possible computations does terminate, the situation is treated as "undefined", which means that the set of results is empty. Using this approach, there is no need for a special symbol $\perp$. [Hansoul 83] uses this approach, because of its technical advantages. However, from a semantic point of view, it is somehow confusing that the representative value for "error" (i.e. $\emptyset$) is contained within any result set (since it is contained in any set, mathematically).

From a model A, according to approach (a), a model C, according to approach (c), can be constructed by defining (for a term t):

$$I^C[t] = I^A[t] \setminus \{\bot\}.$$

The approach (a) is the most general one of the alternative approaches. It corresponds well to the idea that the nondeterministic decision is made *locally* within the operation f. Therefore, the following text develops approach (a) to more detail. In analogy to section 1.1.2, the erratic approach describes the input-output behaviour of an operational unit

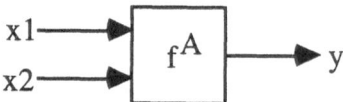

by observations like:

> "If the input lines have the (defined) values x1 and x2, then the output line may have the value y (as one out of all possibilities)."

and

> "If the input lines have the (defined) values x1 and x2, the computation may not terminate (as one out of all possibilities)."

6.1.2 Partial Multi-Algebras

In the following, the notions which have been defined for nondeterministic specifications are generalized to the case of partial operations.

Definition 6.1 (Partial Σ-Multi-Algebra)

Let $\Sigma = (S, F)$ be a signature. A *partial Σ-multi-algebra* A is a tuple A $= (S^A, F^A)$, consisting of

- a family S^A of non-empty carrier sets
  $$S^A = (s^A)_{s \in S}, \quad s^A \neq \emptyset \text{ for } s \in S$$

- a family F^A of set-valued functions
  $$F^A = (f^A)_{f \in F}$$

such that for $[f: s_1 \times \ldots \times s_n \to s] \in F$:

$$f^A: s_1{}^A \times \ldots \times s_n{}^A \to \wp^+((s^A)^\perp)$$

The class of all partial Σ-multi-algebras is called $PAlg(\Sigma)$.

$PGen(\Sigma)$ denotes the term-generated algebras out of $PAlg(\Sigma)$ (according to definition 2.21). $\qquad\qquad\qquad\qquad\qquad\qquad\qquad\qquad\qquad\qquad\qquad\lozenge$

In accordance with the strictness requirement for all operations, the notion of a environment remains unchanged. Environments do not assign a variable to the pseudo-value $\perp$.

Definition 6.2 (Interpretation in Partial Algebras)

Let A be a Σ-algebra, β an environment of X in A.

The *interpretation* $I_\beta^A = (I_{\beta,s}^A)_{s \in S}$ is given by (where $s \in S$):

$$I_{\beta,s}^A: W(\Sigma, X)_s \to \wp^+((s^A)^\perp)$$

using the inductive definition:

(1) If $t = x$ and $x \in X_s$:

$$I_{\beta,s}^A[t] = \{\ \beta(x)\ \}$$

(2) If $t = f(t_1, \ldots, t_n)$ such that $[f: s_1 \times \ldots \times s_n \to s] \in F$:

$$I_{\beta,s}^A[f(t_1,\ldots,t_n)] = \{\ e \in f^A(e_1,\ldots,e_n)\ |\ e_i \in I_{\beta,s_i}^A[t_i] \setminus \{\perp\}\}$$

$$\cup\ \{\ \perp\ |\ \exists\ i \in \{1,\ldots,n\}:\ \perp \in I_{\beta,s_i}^A[t_i]\ \}\qquad\lozenge$$

The notion of interpretation now covers (besides the additive extension) also the strict extension of the operations to undefined values.

In order to create a specification language which excludes models where all operations are completely undefined, an additional kind of axioms is introduced. The *definedness predicate* DEF (compare [Broy, Wirsing 82]) is used to specify whether the interpretation of a term is required to be defined.

Definition 6.3 (DEF-Axiom, Validity)

A *(Σ, X-) DEF-axiom* is a term, written as a formula:

DEF(t) $\qquad\qquad$ where $t \in W(\Sigma, X)$.

For $A \in PAlg(\Sigma)$ we define:

$$A \models DEF(t)$$

iff for all environments $\beta \in ENV(X,A)$: $\perp \notin I_\beta^A[t]$.

The validity of DET-axioms and inclusion rules remains as it was defined in definition 1.8 and 2.2, respectively.   ◊

Please note that this definition implicitly made a decision for the "strong" interpretation of the $\rightarrow$-relation, in analogy to [Broy, Wirsing 82]. For instance if in A we have:

$$f^A = \{ \perp, a \}, \quad g^A = \{ \perp \},$$

then the both formulae hold:

$$A \models f \rightarrow g \quad \text{and} \quad A \models DET(g) .$$

Alternatively, an "existential" interpretation of $\rightarrow$ is possible (analoguously to the so-called "existential equality"), which could be defined by

$$A \models t1 \rightarrow t2 \iff \forall \beta \in ENV(X,A): \perp \notin I_\beta^A[t1] \wedge I_\beta^A[t1] \supseteq I_\beta^A[t2] .$$

The notion of an algebraic specification is extended from now on in such a way that DEF-axioms are admitted as axioms. The class PMod(T) is the class of all partial multi-algebras from $PAlg(\Sigma)$, which are a model of the specification $T = (\Sigma, R)$, i.e. where all axioms $\phi \in R$ are valid. PGen(T) denotes the term-generated algebras in PMod(T).

Example 6.4

A well-known example ([Subrahmanyam 81]) for partial operations is constituted by the structure of finite sets over a basic sort, together with a nondeterministic choice operation:

spec SET
basedon ELEM   { contains the basic sort El (for "elements") }
sort   El, Set
func   empty: $\rightarrow$ Set
            insert: Set $\times$ El $\rightarrow$ Set
            choose: Set $\rightarrow$ El
axioms
            DEF(empty),   DET(empty),

$\qquad$ DEF(insert(s,x)),   DET(insert(s,x)),

$\qquad$ DEF(choose(insert(s,x))),

$\qquad$ insert(insert(s,x),y) $\twoheadrightarrow$ insert(insert(s,y),x),

$\qquad$ insert(insert(s,x),x) $\twoheadrightarrow$ insert(s,x),

$\qquad$ choose(insert(s,x)) $\twoheadrightarrow$ x,

$\qquad$ choose(insert(s,x)) $\twoheadrightarrow$ choose(s)

end

A model S of SET is given by:

$set^S = \wp_{fin}(El^S)$,

$empty^S = \{\ \emptyset\ \}$, $insert^S(M,e) = \{\ M \cup \{e\}\ \}$,

$choose^S(M) = \begin{cases} \{\bot\} & \text{if } M = \emptyset \\ M & \text{else.} \end{cases}$ (where $M \subseteq El^S$, $e \in El^S$).

A "non-standard" model of SET is NS:

$set^{NS} = \{\ c\ \}$ (where c is an arbitrary constant),

$empty^{NS} = \{\ c\ \}$, $insert^{NS}(c,e) = \{\ c\ \}$ (for all $e \in El^{NS}$),

$choose^{NS}(c) = El^{NS}$.

Both models of SET are independent of the actual choice of a model for ELEM. Again, we want the model NS to be excluded, since it is not maximally deterministic. In NS the following fact is valid:

$\qquad$ NS $\models$ choose(empty) $\twoheadrightarrow$ t

for an arbitrary defined term $t \in W(\Sigma)_{El}$. $\qquad\qquad\qquad\qquad$ $\Diamond$

6.2 Partiality and Term Rewriting

The literature on term rewriting usually does not address other models than those of classical equational logic (where all operations are total). A calculus for term rewriting with partial operations has not yet been studied explicitly. But there exists a method to build an equational calculus for partial operations on top of the classcial (total) case ([Broy, Pair, Wirsing 84]). Basically, a "call-by-value" evaluation is simulated within the calculus. For this purpose, at every application of an axiom it is taken care that the instances for variables are defined (similarly to the simulation of "call-time-choice" by conditions formula-ed with DET-predicates). The following definitions follow this idea.

It is an obvious disadvantage of this approach that the pure term rewriting calculus is left. Since the original term rewriting calculus admits *arbitrary* terms as the instances of variables (including terms with an undefined interpretation), classical term rewriting can be used as a calculus only if models with *non-strict* operations are admitted. This approach seems to be promising (cf. also [Broy 87]); but it cannot be worked out within this text.

6.2.1 A Calculus for Partial Specifications

A suitable calculus for partial specifications can be constructed in analogy to definition 2.4. New ingredients are additional preconditions for the instantiation of variables in axioms and particular rules for the DEF-predicate (including strictness rules):

Definition 6.5 (Term Rewriting Calculus with DEF and DET)

> Let $T = (\Sigma, R)$ be a specification with DET- and DEF-axioms.
> A formula $\langle t1 \to t2 \rangle$, $\langle DET(t) \rangle$, or $\langle DEF(t) \rangle$, respectively, is called *deducible* in T, symbolically written
>
> $$T \vdash t1 \to t2, \; T \vdash DET(t), \text{ or } T \vdash DEF(t),$$
>
> iff there is a deduction using the following deduction rules:
>
> (REFL), (TRANS), (CONG), (DET-X), (DET-D), (DET-R)
> $$\text{as in definition 2.4}$$
>
> (AXIOM-1-D)
>
> $$\frac{DET(\sigma x_1), ..., DET(\sigma x_n), DEF(\sigma x_1), ..., DEF(\sigma x_n)}{\sigma l \to \sigma r}$$
>
> $$\text{if } \langle l \to r \rangle \in R, \sigma \in SUBST(\Sigma, X),$$
> $$\{x_1, ..., x_n\} = Vars(l) \cup Vars(r)$$

(AXIOM-2-D)

$$\frac{\text{DET}(\sigma x_1), \ldots, \text{DET}(\sigma x_n), \text{DEF}(\sigma x_1), \ldots, \text{DEF}(\sigma x_n)}{\text{DET}(\sigma t)}$$

if $\langle\text{DET}(t)\rangle \in R, \sigma \in \text{SUBST}(\Sigma, X),$
$\{x_1, \ldots, x_n\} = \text{Vars}(t)$

(AXIOM-3-D)

$$\frac{\text{DET}(\sigma x_1), \ldots, \text{DET}(\sigma x_n), \text{DEF}(\sigma x_1), \ldots, \text{DEF}(\sigma x_n)}{\text{DEF}(\sigma t)}$$

if $\langle\text{DEF}(t)\rangle \in R, \sigma \in \text{SUBST}(\Sigma, X),$
$\{x_1, \ldots, x_n\} = \text{Vars}(t)$

(DEF-X) $\dfrac{\quad}{\text{DEF}(x)}$ if $x \in X$

(DEF-D) $\dfrac{\text{DEF}(t1), \quad t1 \to t2}{\text{DEF}(t2)}$ if $t1, t2 \in W(\Sigma, X)$

(STR) $\dfrac{\text{DEF}(f(t_1, \ldots, t_n))}{\text{DEF}(t_i)}$

if $i \in \{1, \ldots, n\}$,
$[f: s_1 \times \ldots \times s_n \to s] \in F,$
$t_j \in W(\Sigma, X)_{s_j}$ for all $j \in \{1, \ldots, n\}$ ◊

Theorem 6.6 (Soundness)

Let $T = (\Sigma, R)$ be a specification with DET- and DEF-axioms. Then for $t, t1, t2 \in W(\Sigma, X)$:

$$T \vdash t1 \to t2 \quad \Rightarrow \quad \text{PMod}(T) \models t1 \to t2$$
$$T \vdash \text{DET}(t) \quad \Rightarrow \quad \text{PMod}(T) \models \text{DET}(t)$$
$$T \vdash \text{DEF}(t) \quad \Rightarrow \quad \text{PMod}(T) \models \text{DEF}(t).$$

Proof: Analoguously to theorem 2.6, see appendix A.   ◊

6.2.2 Partial DET-Completeness and DET-Additivity

In order to generalize the techniques which have been used in chapter 2, DET-completeness and DET-additivity have to be defined for partial specifications. In the following, again in analogy to [Broy, Wirsing 82], the existence of a deterministic $\rightarrow$-successor term is required only for provably defined terms. The initial model will interpret any term as undefined which cannot be reduced to a provably defined term.

Definition 6.7 (Partial DET-Completeness and DET-Additivity)

Let $T = (\Sigma, R)$ be a specification with DEF- and DET-axioms.

T is called *partially DET-complete* iff:
$$\forall\, t \in W(\Sigma):\ T \vdash DEF(t) \ \Rightarrow$$
$$\exists\, t' \in W(\Sigma):\ T \vdash t \rightarrow t' \ \wedge\ T \vdash DET(t')\ .$$

A term $t \in W(\Sigma)$ is called *potentially undefined* (symbolically: $T \vdash \uparrow t$) iff:
$$\not\exists\, t' \in W(\Sigma):\ T \vdash t \rightarrow t' \ \wedge\ T \vdash DEF(t')\ .$$

T is called *partially DET-additive* iff the following conditions (1) and (2) are fulfilled:

(1) $\forall [f: s_1 \times \ldots \times s_n \rightarrow s] \in F:$
     $\forall\, t_1 \in W(\Sigma)_{s_1}, \ldots, t_n \in W(\Sigma)_{s_n}, t \in W(\Sigma)_s:$

     $T \vdash f(t_1, \ldots, t_n) \rightarrow t \ \wedge\ T \vdash DET(t) \ \wedge\ T \vdash DEF(t) \ \Rightarrow$
     $\exists\, t_1' \in W(\Sigma)_{s_1}, \ldots, t_n' \in W(\Sigma)_{s_n}:$

     $T \vdash f(t_1', \ldots, t_n') \rightarrow t \ \wedge\ T \vdash t_1 \rightarrow t_1' \ \wedge\ \ldots \ \wedge\ T \vdash t_n \rightarrow t_n' \ \wedge$
     $T \vdash DET(t_1') \ \wedge\ \ldots \ \wedge\ T \vdash DET(t_n') \ \wedge$
     $T \vdash DEF(t_1') \ \wedge\ \ldots \ \wedge\ T \vdash DEF(t_n')$

(2) $\forall [f: s_1 \times \ldots \times s_n \rightarrow s] \in F:$
     $\forall\, t_1 \in W(\Sigma)_{s_1}, \ldots, t_n \in W(\Sigma)_{s_n}, t \in W(\Sigma)_s:$

     $T \vdash f(t_1, \ldots, t_n) \rightarrow t \ \wedge\ T \vdash \uparrow t \ \Rightarrow$

$$\exists\, t_1' \in W(\Sigma)_{s_1}, ..., t_n' \in W(\Sigma)_{s_n}:$$
$$T \vdash f(t_1',...,t_n') \to t \;\wedge\; T \vdash t_1 \to t_1' \;\wedge\; ... \;\wedge\; T \vdash t_n \to t_n' \;\wedge$$
$$T \vdash DET(t_1') \;\wedge\; ... \;\wedge\; T \vdash DET(t_n') \;\wedge$$
$$T \vdash DEF(t_1') \;\wedge\; ... \;\wedge\; T \vdash DEF(t_n') \qquad\qquad \Diamond$$

The formulation of DET-additivity contains a new condition (2) which prescribes an additive behaviour for all operations also with respect to the pseudo-value $\perp$. Analoguous syntactical criteria for DET-additivity, as they have been formulated in theorem 2.11, can guarantee partial DET-additivity including condition (2) (see section 6.3).

In analogy to chapter 2, now a term model can be constructed (which later will turn out to be initial for a particular model class, too). In order to achieve non-empty carrier sets, the notion of a sensible signature is used in a slightly modified sense: From now on, a specification $T = (\Sigma, R)$ with signature $\Sigma = (S, F)$ is called *sensible* iff for every sort there exists at least one defined term:

$$T \text{ is sensible } \Leftrightarrow_{def} \forall\, s \in S: \exists\, t \in W(\Sigma)_s: T \vdash DEF(t)$$

Definition 6.8 (Term Model $P\Sigma/R$)

Let $T = (\Sigma, R)$ be a partially DET-complete and sensible specification. A partial Σ-algebra $P\Sigma/R$ is defined by:

$$s^{P\Sigma/R} = \{[t] \mid t \in W(\Sigma)_s \;\wedge\; T \vdash DET(t) \;\wedge\; T \vdash DEF(t)\}$$
$$\text{for } s \in S,$$
$$f^{P\Sigma/R}: W(\Sigma)_{s_1}/\approx \times ... \times W(\Sigma)_{s_n}/\approx \;\to\; \wp^+((W(\Sigma)_s/\approx)^\perp)$$
$$f^{P\Sigma/R}([t_1],...,[t_n]) =$$
$$\{[t] \mid t \in W(\Sigma)_s \;\wedge\; T \vdash f(t_1,...,t_n) \to t$$
$$\wedge\; T \vdash DET(t) \;\wedge\; T \vdash DEF(t)\}$$
$$\cup\; \{\perp \mid \exists\, t: f(t_1,...,t_n) \to t \;\wedge\; \uparrow t\}$$
$$\text{for } [f: s_1 \times ... \times s_n \to s] \in F.$$

Partial DET-completeness ensures that $f^{P\Sigma/R}([t_1],...,[t_n]) \ne \emptyset$. We have always $s^{P\Sigma/R} \ne \emptyset$, since for every sort there exists at least one defined term. The term equivalence from definition 2.12 is again used here, it is denoted by $\approx$. $\qquad\qquad \Diamond$

Theorem 6.9

For a partially DET-complete, partially DET-additive and sensible specification $T=(\Sigma,R)$, $P\Sigma/R$ is a term-generated model of T.

Proof:

Analoguously to theorem 2.14, see appendix A.   ◊

A consequence of theorem 6.9 is the following weak completeness result for PMod(T):

Corollary 6.10 (Weak Completeness for Ground Terms)

Under the preconditions of theorem 6.9 for t1, $t2 \in W(\Sigma)$ the following holds:
$$T \vdash DET(t2) \wedge T \vdash DEF(t2) \wedge PMod(T) \models t1 \rightarrow t2$$
$$\Rightarrow T \vdash t1 \rightarrow t2 .$$

Proof:

$$PMod(T) \models t1 \rightarrow t2$$
$\Rightarrow$ $\quad P\Sigma/R \models t1 \rightarrow t2$ $\qquad\qquad$ (Theorem 6.9)
$\Rightarrow$ $\quad \forall t': \vdash DET(t') \wedge \vdash DEF(t') \wedge \vdash t2 \rightarrow t' \Rightarrow \vdash t1 \rightarrow t'$
$\qquad\qquad\qquad\qquad\qquad\qquad\qquad\qquad$ (Lemma 6.9.1)
$\Rightarrow$ $\quad T \vdash t1 \rightarrow t2$ $\qquad\qquad\qquad$ (Assumptions, (REFL)). ◊

6.3 Partial Specifications with Constructor Basis

At first sight, the language and calculus for partial nondeterministic specifications look rather clumsy and difficult to use. However, if the theory of partial nondeterministic specifications is combined with constructor-based specifications, this language can be made not only more expressive, but also simpler in some sense. The simplification consists in removing the restriction to constructor-completeness of the axiom set, and it has already been used within chapter 4 (sections 4.4.1 ff.) Using the technical machinery from above, here the details are given which justify the simplification.

For this purpose, definition 4.13 for constructor-based specifications is extended
as follows:

Definition 6.11 (Partial Constructor-Based Specification)

Definition 4.13 is extended by the following convention, in order to
interpret a constructor-based specification as an abbreviation for a
specification with DET- and DEF-axioms:

(3)   R does not contain DEF-axioms. All models of T implicitly must
fulfill the following axioms:
$$DEF(c(x_1,...,x_n))$$
for all constructors $c \in C$ (where $x_1, ...,x_n$ are pairwise distinct
fresh variables).   ◊

It is a significant simplification compared to general DET- and DEF-axioms that
the constructor terms are exactly those terms for which definedness and
determinacy can be proven:
$$\forall t \in W(\Sigma): t \in W(\Sigma_C) \Leftrightarrow T \vdash DEF(t) \wedge T \vdash DET(t) \qquad (*)$$

Also in comparison to total constructor-based specifications a simplification is
achieved: Analoguously to the criterion for partial sufficient completeness from
[Broy, Wirsing 82] (equations in "output-normal form"), for constructor-based
specifications the check for C-completeness (complete case analysis over
constructor terms) is unnecessary.

Theorem 6.12

Every partial constructor-based specification is partially DET-complete
and partially DET-additive.

Proof:

The partial additivity follows from theorem 2.11 again. From the proof
of theorem 2.11 it can be seen that the additional condition (2) in
definition 6.7 is fulfilled, too.
The partial DET-completeness follows from the fact that the
definedness can be proven only for constructor terms, and (according to
the definition) exactly for these terms the determinacy can be proven. ◊

Theorem 4.19 (on hierarchical specifications) can be generalized to the case of partial specifications, if a suitable generalization of sufficient completeness to "partial sufficient completeness" (analoguously to [Broy, Wirsing 82]) is used. Under the syntactical precondition mentioned in theorem 4.19, partial constructor-based specifications automatically are partially sufficiently complete and hierarchy-consistent.

The calculus of constructor-based term rewriting from definition 4.20 is obviously sound also for partial specifications, since (*) exactly gives the necessary preconditions to transform any application of (AXIOM-1-C) (see definition 4.20) into one of (AXIOM-1-D) (see definition 6.5). All the deduction rules in definition 6.5 dealing with the deduction of DET- and DEF-formulae are unnecessary in constructor-based specifications; they are replaced by (*).

Combining corollary 6.10 with (*), we immediately have the weak completeness result for partial constructor-based specifications:

$$\forall\ t1 \in W(\Sigma), t2 \in W(\Sigma_C):\ PMod(T) \models t1 \rightarrow t2 \quad \Rightarrow \quad T \vdash_C t1 \rightarrow t2.$$

Constructor-based specifications are sufficiently expressive for the description of many nondeterministic functions which are relevant in practice (cf. chapters 5 and 7). Despite of this power, they are semantically simple: No additional conditions have to be checked in order to provide a well-defined operational and mathematical semantics for them. As it has been shown in chapter 4, they are closely connected with definite logic programs. If we compare the simple calculus of SLD-resolution with the calculus from definition 6.5 above, it becomes evident that logic programming picks a special case out of rather complex surroundings, which gives a good compromise between technical simplicity and expressive power.

Due to the results from chapter 4, it is obvious that a translation into Prolog, or an implementation by graph rewriting are well-suited for experiments with partial constructor-based specifications. The removal of the constructor-completeness condition, however, has the consequence that innermost rewriting is no longer a sound implementation technique. The following example illustrates a case where constructor-based term rewriting differs from innermost term rewriting.

Example 6.13

> **spec** STACK
> **basedon** NAT
> **sort** Stack
> **cons** empty: $\rightarrow$ Stack,
> append: Stack × Nat $\rightarrow$ Stack
> **func** first: Stack $\rightarrow$ Nat
> rest: Stack $\rightarrow$ Stack
> **axioms**
> first(append(s,x)) $\rightarrow$ x,
> rest(append(s,x)) $\rightarrow$ s
> **end**

There are models A∈PMod(STACK) such that the following holds:

$$\text{first}^A(\text{empty}^A) = \{\bot\}, \quad \text{rest}^A(\text{empty}^A) = \{\bot\},$$
$$I^A[\text{rest}(\text{append}(\text{first}(\text{empty}),\text{empty}))] = \{\bot\}.$$

(For an example of such a model, confer PΣ/STACK.)
Constructor-based term rewriting respects this definition. So the term

rest(append(first(empty),empty))

is in normal form with respect to ⊦-$_C$ (i.e. it cannot be reduced to constructor form, it is "undefined" in PΣ/STACK). Innermost term rewriting, however, performs the following, unsound, computation:

$$\text{rest}(\text{append}(\text{first}(\text{empty}),\text{empty})) \; \xrightarrow[\text{STACK}]{\text{im}} \; \text{empty} \; . \qquad \Diamond$$

At this point, the study of nondeterminism in algebraic specifications, term rewriting and algebraic programming has reached a stage of completeness. Partial nondeterministic specifications provide a rich and powerful framework, where classical deductive frameworks can be identified as simple special cases.

Chapter 7 below concludes this text with a larger case study. In order to complete the study also from the semantical point of view, section 6.4 has been included, which extends the results on the structure of model classes from chapter 3 to the partial case. This section addresses only readers interested in model-theoretic semantics of algebraic specifications. For readers interested mainly in the deductive aspects of the framework, it is recommended to skip directly to chapter 7 from here.

6.4 Structure of the Model Classes

This section generalizes the results on the structure of model classes from chapter 3 (in particular existence of initial and terminal models) to the partial case. This leads to a rather complex theory, which is mainly burdened with many slightly different notions. Below, the notions are chosen in such a way that the proof techniques from chapter 3 can be easily carried over. However, there may be various ways of fine-tuning the definitions in these respects.

An important and not easy topic is the determination of a suitable notion of homomorphism for partial nondeterministic algebras. For partial deterministic algebras appropriate notions have been developed in [Broy, Wirsing 82]: This work distinguishes between weak, total and strong homomorphisms. In the following, this work is generalized to the case of set-valued functions. This leads to an even larger number of different notions, since the distinction between loose and tight homomorphism must be combined with all three different homomorphisms. Fortunately, for initiality results only some particular combinations are of interest.

6.4.1 Homomorphisms

Definition 6.14 (Σ-Homomorphisms for Partial Algebras)

Let $\Sigma = (S, F)$ be a signature, $A, B \in PAlg(\Sigma)$.
A *loose Σ-homomorphism* φ from A to B is the strict continuation of a family of mappings

$$\varphi = (\varphi_s)_{s \in S}, \qquad \varphi_s \colon s^A \rightarrow \wp^+((s^B)^\perp),$$

which fulfils the following condition:

For all $[f: s_1 \times \ldots \times s_n \rightarrow s] \in F$ and all $e_1 \in s_1^A, \ldots, e_n \in s_n^A$:
$$\{e' \in \varphi_s(e) \mid e \in f^A(e_1, \ldots, e_n) \wedge e \neq \perp\}$$
$$\subseteq \{e' \in f^B(e_1', \ldots, e_n') \mid e_1' \in \varphi_{s_1}(e_1), \ldots, e_n' \in \varphi_{s_n}(e_n)\}$$

φ is called a *tight Σ-homomorphism*, iff the stronger condition holds for all $e_1, \ldots, e_n$:

$$(\perp \notin f^A(e_1, ..., e_n) \Rightarrow$$
$$\{ e' \in \varphi_s(e) \mid e \in f^A(e_1,...,e_n) \}$$
$$= \{ e' \in f^B(e_1',...,e_n') \mid e_i' \in \varphi_{s_i}(e_i) \})$$
$$\wedge \quad (\perp \in f^A(e_1, ..., e_n) \Rightarrow$$
$$\{ e' \in \varphi_s(e) \mid e \in f^A(e_1,...,e_n) \wedge e \neq \perp \}$$
$$= \{ e' \in f^B(e_1',...,e_n') \mid e_i' \in \varphi_{s_i}(e_i) \wedge e' \neq \perp \})$$

φ is called *elementary*, iff $\forall\ e \in s^A$: $|\varphi(e)| = 1$.

φ is called *total*, as usual, iff $\forall\ e \in s^A$: $\perp \notin \varphi(e)$.

φ is called *weak*, iff for all $x_1, ..., x_n$:

$$\perp \in f^A(e_1,...,e_n) \Rightarrow$$
$$\exists\ e_1' \in \varphi_{s_1}(e_1), ..., e_n' \in \varphi_{s_n}(e_n): \perp \in f^B(e_1', ..., e_n').$$

φ is called *strong*, iff φ is total and weak.

A partial multi-algebra A is called *loosely initial* in a class K of partial multi-algebras, iff for all $B \in K$ there exists a unique total and loose Σ-homomorphism $\varphi: A \rightarrow B$. A is called *strongly initial*, iff for all $B \in K$ there exists a unique strong and tight Σ-homomorphism $\varphi: A \rightarrow B$.

A partial multi-algebra A is called *weakly terminal* in a class K of partial multialgebras, iff for all $B \in K$ there exists an elementary, weak and loose Σ-homomorphism $\varphi: B \rightarrow A$. A is called *strongly terminal*, iff for all $B \in K$ there exists an elementary, strong and tight Σ-homomorphism $\varphi: B \rightarrow A$. ◊

Similar definitions can be found in [Nipkow 86], but only for the combinations loose/total and tight/strong (and with different naming conventions). The notion of homomorphism in [Hansoul 83] is similar to tight homomorphisms, but for a simpler notion of a partial multi-algebra (approach (c) according to section 6.1.1). The notion of weak terminality has been introduced here only in order to illustrate similarities and differences to [Broy, Wirsing 82].

The rather complex definition of a tight homomorphism is motivated by the fact that this definition leads to simpler notions when it is combined with the property "total" or "weak". This can be seen best for term-generated algebras.

Lemma 6.15

Let Σ be a signature, A, B$\in$PGen(Σ), φ: A $\to$ B a loose Σ-homomorphism. Then:

(1) φ is total $\Leftrightarrow$
      $\forall$ t$\in$W(Σ): $\{$ e'$\in\varphi$(e) | e$\in$I^A[t]\\$\{\perp\}\} \subseteq$ I^B[t]\\$\{\perp\}$

(2) φ is tight and total $\Leftrightarrow$
      $\forall$ t$\in$W(Σ): $\{$ e'$\in\varphi$(e) | e$\in$I^A[t]\\$\{\perp\}\}$ = I^B[t]\\$\{\perp\}$
                    $\wedge$ ($\perp\in$I^B[t] $\Rightarrow$ $\perp\in$I^A[t])

(3) φ is weak $\Leftrightarrow$
      $\forall$ t$\in$W(Σ): $\{$ e'$\in\varphi$(e) | e$\in$I^A[t] $\} \subseteq$ I^B[t]

(4) φ is tight and weak $\Leftrightarrow$
      $\forall$ t$\in$W(Σ): $\{$ e'$\in\varphi$(e) | e$\in$I^A[t] $\}$ = I^B[t]

(5) φ is tight and strong $\Leftrightarrow$
      $\forall$ t$\in$W(Σ): $\{$ e'$\in\varphi$(e) | e$\in$I^A[t] $\}$ = I^B[t]
                    $\wedge$ ($\perp\in$I^A[t] $\Leftrightarrow$ $\perp\in$I^B[t])

Proof: See appendix A. $\Diamond$

6.4.2 Initial Algebras

Under the preconditions of theorem 6.9, a loose Σ-homomorphism from PΣ/R into an arbitrary model A$\in$PMod(T) exists (just take the continuation of the interpretation I^A). Therefore we have:

Theorem 6.16

If T = (Σ, R) is partially DET-complete, partially DET-additive and sensible, PΣ/R is loosely initial in PMod(T) and PGen(T).

Proof:

The proof is conducted in analogy to theorem 3.13 (see appendix A).
For [t]$\in$s$^{P\Sigma/R}$ holds: |- DEF(t), therefore $\perp\notin$I^A[t]. This means that the continuation of I^A is a total Σ-homomorphism from PΣ/R to A. $\Diamond$

In order to state connections to the notions of homomorphism used in the literature, a similar result to lemma 3.13 was useful. But the generalization of the lemma would require that for an arbitrary model A$\in$PGen(T) the following property always holds (which is not the case):

$$\forall\ e \in s^A: \exists\ t \in W(\Sigma):\ T \vdash DEF(t)\ \wedge\ I^A[t] = \{e\}$$

A particular class of models is characterized by this property.

Definition 6.17 (Minimally Defined Models)

Let Σ be a signature, $K \subseteq PAlg(\Sigma)$ a class of algebras, $A \in PAlg(\Sigma)$.

A term $t \in W(\Sigma)$ is called *undefined* in A (symbolically: $A \models \uparrow t$) iff:

$$I^A[t] = \{\bot\}.$$

A is called *minimally defined* in K iff:

$$\forall\ t \in W(\Sigma): (\exists\ B \in K: B \models \uparrow t)\ \Rightarrow\ A \models \uparrow t\ .$$

Lemma 6.18

Let $T = (\Sigma, R)$ be a partially DET-complete, partially DET-additive and sensible specification.

(1)   For $A \in PMod(T)$:

A is minimally defined in PMod(T) $\Leftrightarrow$

$$(\forall\ t \in W(\Sigma):\ A \models \uparrow t\ \Leftrightarrow\ T \vdash \uparrow t\ )$$

(2)   If a term-generated model $A \in PGen(T)$ is minimally defined in PMod(T), then:

$$\forall\ e \in s^A: \exists\ t \in W(\Sigma):\ T \vdash DEF(t)\ \wedge\ I^A[t] = \{e\}$$

(3)   Let $A \in PGen(T)$, $B \in PMod(T)$, A minimally defined in PMod(T). Then all loose and total homomorphisms $\varphi: A \rightarrow B$ are elementary.

Proof: See appendix A. $\Diamond$

Part (1) of this lemma states that $P\Sigma/R$ is minimally defined in PMod(T), part (3) therefore corresponds to lemma 3.12. As a consequence, only elementary homomorphisms appear in the initiality results below.

Please note that the notion of minimal definedness from above is *not* identical to the notion with the same name in [Broy, Wirsing 82]. The reason for this is that in a nondeterministic framework from the proposition "t is not defined in A" ($\neg(A \models DEF(t))$ we cannot conclude that "t is undefined in A" ($A \models \uparrow t$). The interpretation of t in A may contain defined values *and* the pseudo-value $\perp$. It is perfectly adissible that for a term t we have the following:

$P\Sigma/R \models DEF(t) \quad \wedge \quad \exists A \in PGen(T): \neg(A \models DEF(t))$.

Example 6.19

> **spec** SOME
> **sort** Nat
> **func** zero: $\rightarrow$ Nat,   succ: Nat $\rightarrow$ Nat,
>   some: $\rightarrow$ Nat
> **axioms**
>   DET(zero),   DET(succ(x)),
>   DEF(zero),   DEF(succ(x)),
>   some $\rightarrow$ zero,   some $\rightarrow$ succ(some)
> **end**

The model $P\Sigma/SOME$ is isomorphic to the following model P:
$$Nat^P = \mathbb{N}, \qquad zero^P = \{0\}, \qquad succ^P(n) = \{n+1\},$$
$$some^P = \mathbb{N} .$$

Another model A is given by:
$$Nat^A = \mathbb{N}, \qquad zero^A = \{0\}, \qquad succ^A(n) = \{n+1\},$$
$$some^P = \mathbb{N} \cup \{\perp\} .$$

In the notation of [Broy, Wirsing 82], P cannot be called minimally defined (since P $\models$ DEF(some), $\neg(A \models DEF(some))$). Using the definition above, P is minimally defined.

Independent of this question, A lacks the property of maximal determinacy, since the pseudo-value $\perp$ in the interpretation of some represents a form of "superfluous" nondeterminism, which has no foundation in the specification text. $\Diamond$

The example shows that the formal generalization of the notion "maximally deterministic" has to handle the pseudo-value $\perp$ appropriately. The crucial point is here the definition of the relation "more deterministic than" for partial algebras. As a generalization of the definition in the total case, it is obvious how the sets of *defined* values are to be compared (for the interpretation of a certain term in two algebras). In analogy to the notion of partial correctness, this relation is called *"partially more deterministic"*.

The example above shows that another, "total" notion is needed as well, which describes whether P is more deterministic than A. The formal definition relies on lemma 6.15 (1) and (2).

Definition 6.20 (Maximally Deterministic Models in the Partial Case)

Let A, A' be partial Σ-multi-algebras.

A' is called a *refinement* of A, iff there is a loose total Σ-homomorphism $\varphi: A' \to A$.

A' is called *partially more deterministic* than A, iff:
$$\forall t \in W(\Sigma): (\, |\, I^A[t]\backslash\{\bot\}\, |\, \geq\, |\, I^{A'}[t]\backslash\{\bot\}\, |\,)$$

A' is called *(totally) more deterministic* than A, iff:
$$\forall t \in W(\Sigma): (\, |\, I^A[t]\backslash\{\bot\}\, |\, \geq\, |\, I^{A'}[t]\backslash\{\bot\}\, |\,)$$
$$\wedge\, (\, \bot \in I^{A'}[t]\, \Rightarrow\, \bot \in I^A[t]\,).$$

A is called *maximally deterministic*, iff A is totally more deterministic than any refinement of A.

DPGen(T) denotes the class of maximally deterministic term-generated models of a specification T. $\qquad\qquad\Diamond$

There exists a strong connection between the model class DPGen and the "two-phase" fixpoint semantics for nondeterministic programs defined in [Broy 86]. The maximality constraint with respect to the relation "partially more deterministic" corresponds to a designation of "Egli-minimal" algebras, the treatment of the $\bot$-element leads to an additional minimality constraint with respect to the set inclusion ordering.

Lemma 6.21

Let T be a partially DET-complete, partially DET-additive and sensible specification. For any $A \in PGen(T)$ the following propositions are equivalent:

(1)   A is maximally deterministic.

(2) $\forall$ B$\in$PGen(T):

φ: B$\to$A is a loose and total Σ–homomorphism $\Rightarrow$

φ is a tight Σ-homomorphism.

(3) $\forall$ t$\in$W(Σ):

($\forall$ e$\in$I^A[t]: e $\neq$ $\bot$ $\Rightarrow$ $\exists$ t'$\in$W(Σ):

T |- t $\to$ t' $\wedge$ T |- DET(t') $\wedge$ T |- DEF(t') $\wedge$ I^A[t'] = { e }) $\wedge$

($\bot\in$I^A[t] $\Rightarrow$ $\exists$ t'$\in$W(Σ):

T |- t $\to$ t' $\wedge$ T |- $\uparrow$t' $\wedge$ I^A[t'] = {$\bot$})

Proof: See appendix A. $\Diamond$

Theorem 6.22

Let T be a partially DET-complete, partially DET-additive and sensible specification.

Then PΣ/R is strongly initial in DPGen(T).

Proof:

Let A$\in$DPGen(T). According to theorem 6.16, I^A is a unique loose and total homomorphism from PΣ/R to A. With lemma 6.21 (2), I^A is a tight homomorphism, too.

Let t$\in$W(Σ), $\bot\in$I$^{P\Sigma/R}$[t], then $\exists$ t': T |- t $\to$ t' $\wedge$ T |- $\uparrow$t' (lemma 6.9.1). If it was the case that e$\in$I^A[t'], e$\neq\bot$, then Lemma 6.21 (3) would deliver a contradiction to T |- $\uparrow$t'. Therefore I^A[t'] = {$\bot$}, i.e. $\bot\in$I^A[t]. According to lemma 6.15 (5) this means that I^A is a strong homomorphism, too. $\Diamond$

From theorem 6.22 and lemma 6.18 (1) it follows that all models in DPGen(T) are minimally defined in PMod(T) (in the sense of definition 6.17).

A graphical visualization of the lattice structure of the model classes can be found at the end of the next section.

6.4.3 Terminal Algebras

In the case of partial specifications, the existence of terminal algebras is a less trivial question than in the total case. From the definition of weak homomorphisms it can be seen directly that a (weakly or strong) terminal algebra A in a class of algebras K has to fulfil the following property ("minimal definedness" in [Broy, Wirsing 82]):

$$\forall t \in W(\Sigma): (\exists B \in K: \bot \in I^B[t]) \Rightarrow \bot \in I^A[t]. \tag{MD}$$

This property does not hold for a trivial algebra construction (like $Z\Sigma$ from definition 3.6). Even one-element carrier sets and partial functions do not suffice in general to construct a terminal algebra, since partial specifications implicitly may exclude some possible identifications of elements:

Example 6.23

> **spec** NT
> **sort** s
> **func** a: $\to$ s, b: $\to$ s,   f: s $\to$ s,   g: $\to$ s
> **axioms**
>   DET(a),   DEF(a),   DET(b),   DEF(b),
>   DEF(f(a)),   f(x) $\to$ a,   f(b) $\to$ g
> **end**

> There does not exist a term-generated model of NT, in which a und b are identified and in which the property (MD) holds:

> Let $A \in PGen(NT)$, $a \in s^A$, $a^A = b^A = \{ a \}$.
> Because of $\vdash DEF(f(a))$ then $\bot \notin f^A(a)$.
> In $P\Sigma/NT \in PGen(NT)$ we have: $\bot \in I^{P\Sigma/NT}[f(b)]$, therefore A does not fulfil (MD)   (otherwise from $a^A = b^A$ it followed that $\bot \in f^A(a)$, contradiction). $\Diamond$

Another negative statement can be deduced from the example above:

Theorem 6.24

> In general, within PGen(T) strong terminal algebras do not exist.

Proof:

Consider the specification NT from example 6.23. Two models A and B of NT are given by:

$$s^A = s^B = \{ a, b \}, \qquad a^A = a^B = \{ a \}, \ b^A = b^B = \{ b \},$$
$$f^A(a) = f^B(a) = \{ a \}, \qquad f^A(b) = f^B(b) = \{ a, \perp \},$$
$$g^A = \{ a, \perp \}, \qquad g^B = \{ b, \perp \}.$$

A and B fulfil (MD). If there was a model Z, as well as tight and strong homomorphisms φ: A $\rightarrow$ Z and ψ: B $\rightarrow$ Z, then according to lemma 6.15 (5) the following statements would be valid:

$$\{ \varphi(a), \perp \} = g^Z = \{ \psi(b), \perp \},$$
$$\{ \varphi(a) \} = a^Z, \quad \{ \psi(b) \} = b^Z.$$

From this follows that $a^Z = b^Z$. As in example 6.23, Z does not fulfil (MD), so it cannot be strongly terminal in PGen(T).   ◊

However, a weakly terminal model for PMod(T) can be constructed. In this model the operations are in general weaker defined and less deterministic than those in models of DPGen(T), so the weakly terminal model is not very interesting for nondeterministic specifications. This is the reason why weakly terminal models in PMod(T) are not studied here. It is a more interesting observation that in the most interesting model class DPGen(T) a strongly terminal model does exist. For the construction of this model we use a relation, which is defined in analogy to [Broy, Wirsing 82].

Definition 6.25

On $W(\Sigma)$ a quasi-ordering $\longrightarrow$ is defined by $(t, t' \in W(\Sigma))$

$$t \longrightarrow t' \quad \Leftrightarrow_{def}$$
$$\exists A_1, ..., A_{k+1} \in DPGen(T), t_1, ..., t_k \in W(\Sigma):$$
$$A_1 \models t \rightarrow t_1 \wedge A_2 \models t \rightarrow t_2 \wedge ... \wedge A_{k+1} \models t_k \rightarrow t'.$$

$\longleftrightarrow$ denotes the equivalence relation induced by $\longrightarrow$:

$$t \longleftrightarrow t' \quad \Leftrightarrow_{def} \quad t \longrightarrow t' \wedge t' \longrightarrow t.$$ ◊

It can be seen easily that $\longleftrightarrow$ is reflexive, transitive und congruent with respect to the term building operations. $\longrightarrow$ induces an ordering on the $\longleftrightarrow$-equivalence classes in $W(\Sigma)$.

Definition 6.26

Let $T = (\Sigma, R)$ be a partially DET-complete, partially DET-additive and sensible specification.

An extended axiom set $E(R)$ is defined by

$E(R) = R \cup \{ \langle t \rightarrow t' \rangle \mid t, t' \in W(\Sigma) \land$

$\vdash DET(t) \land \vdash DEF(t) \land \vdash DET(t) \land \vdash DEF(t) \land t \Leftrightarrow t' \}.$

Obviously, $E(R)$ is partially DET-complete and DET-additive, again.

Therefore the algebra $P(\Sigma)/E(R)$ is well-defined. $\Diamond$

Theorem 6.27

Under the preconditions of definition 6.26, $P\Sigma/E(R)$ is strongly terminal in DPGen(T).

Proof: See appendix A. $\Diamond$

For the specification NT from example 6.23 the strongly initial and the strongly terminal algebra in DPGen(T) are isomorphic. Essentially only one algebra is specified here ("monomorphic specification").

Altogether, the structure of the model class of a partially DET-complete, partially DET-additive and sensible specification $T = (\Sigma, R)$ can be visualized graphically as follows:

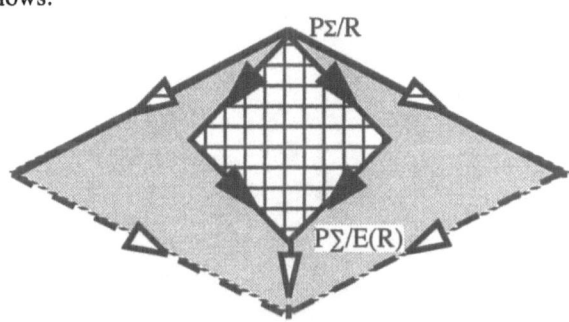

PΣ/R

PΣ/E(R)

 tight and strong homomorphisms

 loose and total homomorphisms

 loose and weak homomorphisms

 PMod(T)

 DPGen(T)

Chapter 7

Communicating Processes: An Example

In this chapter the theory developed above shall be applied to a non-trivial example, in order to demonstrate that the original aims are fulfilled. Moreover, the example leads to some ideas how the current state of development can be improved.

The specification presented in this chapter has been tested (with only a few modifications) using the RAP system. Some of the experiments are documented in appendix B.

As an example, the operational semantics of a programming language for communicating sequential processes is specified. This example is of particular interest, since one of the main motivations for the treatment of nondeterminism comes from the field of programming parllel and communicating processes. The language chosen here is a simplified variant of the language CSP ([Hoare 78]). Syntax and semantics have been taken from [Olderog, Hoare 86], except of minimal notational differences. A very similar language constitutes the basis for the concepts of [Broy 84] .

7.1. Communicating Processes (CP)

The language of Communicating Processes (CP) presupposes a given alphabet C of elementary actions (*communication actions*).

The syntax of CP-programs in Backus-Naur form is as follows:

⟨Agent⟩ ::=   **stop** |
              div |
              ⟨Action⟩ → ⟨Agent⟩ |
              ⟨Agent⟩ **OR** ⟨Agent⟩ |
              ⟨Agent⟩ [] ⟨Agent⟩ |
              ⟨Agent⟩ ‖ {⟨ActionSet⟩} ⟨Agent⟩ |
              ⟨Id⟩ :: ⟨Agent⟩ |
              ⟨Id⟩

Here ⟨Action⟩ denotes elements from C, ⟨ActionSet⟩ means finite sets of elements from C and ⟨Id⟩ denotes identifiers.

The informal meaning of agents is given by the *processes* they describe. A process basically can be understood as a sequence (or, if parallelism is involved, as a partial order) of communication *events*, each of which is an instance of some action out of the set C.

stop

> describes a process which has reached its termination already and is no longer able to communicate. Sometimes this situation occurs within a system of processes unintentionally; then it is called a *deadlock*.

div

> also describes a process which is unable to further communication. But here this failure is due to an infinite sequence of internal computation steps: The process *diverges*.

a → p

> describes the process which is able to perform the action a and which afterwards behaves like the process p. An action in general means a communication with another process.

pORq

> describes a process which behaves nondeterministically either like p or like q. This form of nondeterminism is called *internal nondeterminism*, since the decision is taken by the process itself, completely independently of its environment.

p[]q

> also describes a process which may behave nondeterministically like p
> or q. But here the decision can be controlled by the environment
> (which consists of other processes). The process is only allowed to
> choose such an alternative, if its first action leads to a successful
> communication with its environment. Therefore this form of
> nondeterminism is called *external*.

p‖{A}q

> is used to compose systems of parallel processes. The action set A
> describes the actions for which a *synchronization* of p and q is
> necessary (*internal communication* between p and q).

p \ a

> is used to "hide" internal actions performed by the process p. If the
> process described by p performs the action a, this fact is no longer
> abservable from the outside. The "hiding" construct admits a modular
> construction of more abstract processes from primitive ones.

x :: p and x

> (where x is an identifier) are used to declare processes recursively.

In the following, some approaches to a semantics of CP are specified. The
notational difference between agents (programs) and processes (evaluations of
programs) is handled less rigidly from now on.

7.2. Semantics of CP

In [Olderog, Hoare 86] a number of mathematical models for CP-processes are
given. It was principally possible to describe those semantical models directly
by nondeterministic specifications, since all these models use additive operations
(besides a few problems with hiding, see below).

But in order to stress the relationship to executable specifications and term
rewriting, the specification below uses the *operational* semantics of CP as its
starting point. It turns out that the language of nondeterministic algebraic

specifications suffices to abstract from unnecessary technical details, leading to a rather "abstract" kind of operational semantics.

7.2.1. Transition Semantics

The operational semantics for CP given in [Olderog, Hoare 86] uses *transition systems* (labelled term rewrite rules). A rule in this framework is of the form

$$p \xrightarrow{x} p'$$

where p, p' are agents and x is an action. The rules can be understood as follows: The process described by p is able to perform the action x and then behaves like the process described by p'. The transition semantics admits here a particular psudo-action $\tau \notin C$ in additon to the communication actions (similarly to the ε-transitions in nondeterministic automata).

Transition systems are a rather general tool for the definition of operational semantics, as demonstrated by the "SOS"-style of semantic definition ([Plotkin 81], [Hennessy 90]). Therefore the treatment of transition systems is interesting for a wider area of applications than the actual language studied here.

For an algebraic specification of the semantics, the main problem is an appropriate handling of the labels used in the transition systems. Except of these labels, the concept of non-confluent term rewriting corresponds well to transition systems. In [Meseguer 92] the whole notion of rewriting has been extended to cover the notion of labelled rewriting; we prefer to encode the label associated with a rewriting step. The three basic alternatives for such an encoding are:

(a) Transition rules as predicates

The transition rule is described by a predicate with three arguments:
OP: Agent × Action × Agent → Bool
OP(p,a,p') = true.

This approach is used in [Broy 84]. It means to simulate nondeterminism on a deterministic level (similar to [Subrahmanyam 81]). The models of such a specification do not contain true nondeterministic operations.

(b) Acceptor approach

The transition rule is described by a function

trans: Agent $\times$ Action $\to$ Agent

trans(p,a) $\to$ p',

i. e. the successor state is specified for a given process and a given action.

(c) Generator approach

The transition rule is described by a function

trans: Agent $\to$ Action $\times$ Agent

trans(p) $\to$ ⟨a,p'⟩,

i. e. the successor state and a possible action are specified for a given process.

The difference between variants (b) and (c) can be seen best for an example. In order to compute the successor states of the agent

(a $\to$ **stop**) **OR** (b $\to$ **div**)

the following transition rules from [Olderog, Hoare 86] are needed:

p **OR** q $\xrightarrow{\tau}$ p

p **OR** q $\xrightarrow{\tau}$ q ,

i. e. the transition semantics admits the following possibilities for the first step:

(a $\to$ **stop**) **OR** (b $\to$ **div**) $\xrightarrow{\tau}$ (a $\to$ **stop**)

(a $\to$ **stop**) **OR** (b $\to$ **div**) $\xrightarrow{\tau}$ (b $\to$ **div**)

(This corresponds to the intuitive idea that a process composed by **OR** can decide "spontaneously" in favour of one of its subprocesses.)

A model B following approach (b) models this situation like this:

$$\text{trans}^B(p,x) = \begin{cases} \{a\to\text{\bf stop},b\to\text{\bf div}\} & \text{if } x=\tau \\ \bot & \text{otherwise} \end{cases}$$

A model C according to approach (c) however defines:

$$\text{trans}^C(p) = \{ \langle\tau,a\to\text{\bf stop}\rangle, \langle\tau,b\to\text{\bf div}\rangle \} .$$

Basically both variants are acceptable models of the considered situation. But the second variant mirrors more precisely the intuitive idea of a "spontaneous" action. In approach (c) a process generates the possible τ-actions by itself, whereas in approach (b) the "spontaneous" actions have to be stimulated from the outside by supplying the process with a pseudo-action. Moreover, approach (c) seems to lead to a technically simpler way of modelling. These are the reasons why from now on approach (c) is followed.

In order to give an appropriate semantics for CP, all the concepts developed in the previous chapters must be used (partial specifications, conditional axioms). It is sufficient, however, to consider constructor-based specifications (according to section 4.4). Therefore the specifications do not contain explicit DET- or DEF-axioms; for all constructor terms implicit DET- and DEF-axioms are assumed instead. It is reasonable to construct the specification in a modular (*hierarchical*) way.

The following standard types are used:

BOOL   the truth values, containing the sort Bool and the usual operations;

COM   the basic alphabet for communication actions, containing the sort Com and an equality predicate equal_Com: Com $\times$ Com $\rightharpoonup$ Bool;

ID   the identifiers (strings), containing the sort Id an an equality predicate equal_Id: Id$\times$Id $\rightharpoonup$ Bool .

Actions are described by

spec ACTION
basedon COM
sort   Action
cons   tau: $\rightharpoonup$ Action   { invisible action }
            com: Com $\rightharpoonup$ Action   { communication action }
end

The specification for sets of communication actions is omitted here (see appendix B, specification COM_SET), in order to improve readability standard notation for sets is used here.

The syntax of agents is easily described as an *abstract syntax*:

spec AGENT
basedon ID, COM

sort	Agent	
cons	stop: → Agent	{ **stop** },
	div: → Agent	{ **div** },
	prefix: Com × Agent → Agent	{ a → p },
	OR: Agent × Agent → Agent	{ p **OR** q },
	choice: Agent × Agent → Agent	{ p [] q },
	par: Agent × **SET**(Com) × Agent → Agent	{ p ‖{A} q },
	hide: Agent × Com → Agent	{ p \ a },
	rec: Id × Agent → Agent	{ x :: p },
	call: Id → Agent	{ x }

end

Since all operations in AGENT are syntactical constructors for programs, all the operations are total and deterministic. This includes the nondeterministic constructs OR and choice! The sort Agent describes only the (deterministic and defined) objects of program terms. Nondeterminism or nontermination do not appear unless such programs are executed.

For the definition of a transition semantics, as it was sketched above, tuples

Action × Agent

are needed. Again the mathematical notation is preferred over an explicit specification of these tuples. (The constructor ‹ . , . › for tuples is assumed to be total and deterministic.) For a more detailed version see appendix B (specification PAIR).

The axioms of the specification below are taken directly from [Olderog, Hoare 86], adapted only to the new notational conventions.

spec TRANS
basedon BOOL, ID, ACTION, COM, AGENT, SUBST
func   trans: Agent → **PAIR**(Action,Agent)
axioms

trans(div) → ‹tau,div›,
trans(prefix(i,p)) → ‹com(i),p›,
trans(OR(p,q)) → ‹tau,p›,
trans(OR(p,q)) → ‹tau,q›,
trans(p) → ‹com(i),p'› ⇒ trans(choice(p,q)) → ‹com(i),p'›,
trans(q) → ‹com(i),q'› ⇒ trans(choice(p,q)) → ‹com(i),q'›,
trans(p) → ‹tau,p'› ⇒ trans(choice(p,q)) → ‹tau,choice(p',q)›,

$trans(q) \rightarrow \langle tau,q'\rangle \Rightarrow trans(choice(p,q)) \rightarrow \langle tau,choice(p,q')\rangle$,

$i \in A$ & $trans(p) \rightarrow \langle com(i),p'\rangle$ & $trans(q) \rightarrow \langle com(i),p'\rangle \Rightarrow$
$trans(par(p,A,q)) \rightarrow \langle com(i),par(p',A,q')\rangle$,

$i \notin A$ & $trans(p) \rightarrow \langle com(i),p'\rangle \Rightarrow trans(par(p,A,q)) \rightarrow$
$\langle com(i),par(p',A,q)\rangle$,

$i \notin A$ & $trans(q) \rightarrow \langle com(i),q'\rangle \Rightarrow trans(par(p,A,q)) \rightarrow$
$\langle com(i),par(p,A,q')\rangle$,

$trans(p) \rightarrow \langle tau,p'\rangle \Rightarrow trans(par(p,A,q)) \rightarrow \langle tau,par(p',A,q)\rangle$,

$trans(q) \rightarrow \langle tau,q'\rangle \Rightarrow trans(par(p,A,q)) \rightarrow \langle tau,par(p,A,q')\rangle$,

$trans(p) \rightarrow \langle com(j),p'\rangle$ & $equal\_Com(i,j) = true \Rightarrow$
$trans(hide(p,i)) \rightarrow \langle tau,p'\rangle$,

$trans(p) \rightarrow \langle com(j),p'\rangle$ & $equal\_Com(i,j) = false \Rightarrow$
$trans(hide(p,i)) \rightarrow \langle b,p'\rangle$,

$trans(rec(x,p)) \rightarrow \langle tau,p[x/rec(x,p)]\rangle$

end

Recursion has been treated using a substitution operator:

$\_[\_/\_]$: Agent × Id × Agent $\rightarrow$ Agent

with its usual meaning (for details see appendix B, specification SUBST).

Because of theorem 6.12, TRANS is partially DET-complete and DET-additive.

The operation trans has been specified as a partial function. The following terms, for instance, are interpreted as undefined in DPGen(TRANS):

    $trans(stop)$

    $trans(par(prefix(a,p),\{a,b\},prefix(b,q)))$

    $trans(call(x))$

The first both terms correspond to deadlock situations, the third one contains a "context error" (call of a process name which has not been declared).

The specification TRANS is hierarchy-persistent, as can be seen from the generalization of theorem 4.19 to the partial case. Since TRANS does not introduce any non-primitive sort, an analoguous generalization of theorem 2.33 shows that, once a model for the primitive specifications ACTION, COM, and AGENT has been fixed, the model class DPGen (TRANS) contains essentially one single model. This corresponds well to the idea of having a fixed operational semantics as the basis of further considerations.

7.2.2. Trace Semantics

The τ-transitions used in the transition semantics are operational details, an abstraction from which is interesting. A sequence of actions containing only communication actions is called a *trace*. The following specification describes, in analogy to TRANS, the first possible state transition of a process concerning an action different from τ.

spec STEP0
basedon ACTION, COM, AGENT, TRANS
func   step: Agent → **PAIR**(Com,Agent)
axioms

        trans(p) → ‹com(i),p'› ⟹ step(p) → ‹i,p'›,
        trans(p) → ‹tau,p'› ⟹ step(p) → step(p')

end

Again, STEP0 has essentially one model extending its primitive parts.

The operation step now can be defined now in a simpler way without using the transition semantics. (More formally: STEP0 is an implementation of the following specification STEP.)

spec STEP
basedon AGENT, COM, SUBST, ID
func   step: Agent → **PAIR**(Com,Agent)
axioms

        step(prefix(i,p)) → ‹i,p›,
        step(OR(p,q)) → step(p),
        step(OR(p,q)) → step(q),
        step(choice(p,q)) → step(p),
        step(choice(p,q)) → step(q),
        step(p) → ‹i,p'› & step(q) → ‹i,q'› & i∈A ⟹
            step(par(p,A,q)) → ‹i,par(p',A,q')›,
        step(p) → ‹i,p'› & i∉A ⟹ step(par(p,A,q)) → ‹i,par(p',A,q)›,
        step(q) → ‹i,q'› & i∉A ⟹ step(par(p,A,q)) → ‹i,par(p,A,q')›,
        step(p) → ‹j,p'› & equal_Com(i,j) = true ⟹
            step(hide(p,i)) → step(p'),
        step(p) → ‹j,p'› & equal_Com(i,j) = false ⟹
            step(hide(p,i)) → ‹j,p'›,

        step(rec(x,p)) → step(p[x/rec(x,p)])
end

STEP shows in a rather simple way two typical properties of the trace semantics: Internal and external nondeterminism are treated identically; and no difference is made between **stop** (deadlock) and **div** (divergence).

The specification TRACES is based on STEP. It describes all possible initial segments for all traces of a process:

spec TRACES
basedon COM, AGENT, STEP
sort   Trace
cons   empty: → Trace,
            append: → Com × Trace → Trace
func   trace: Agent → Trace
axioms
        trace(p) → empty,
        step(p) → ⟨i,p'⟩ ⟹ trace(p) → append(i,trace(p'))
end

Please note that TRACES admits non-isomorphic models (since it defines a new sort). The initial model $P\Sigma$/TRACES for divergence-free process exactly corresponds to the *trace model* T in [Olderog, Hoare 86]. More abstract models (for instance corresponding to the *counter model* C) are admitted as models of TRACES, too.

As it was mentioned already, in DPGen(TRACES) there is no difference between **div** and **stop**. A model which corresponds better to the intuitive understanding of the process should have the following properties (call here the model A):
        I^A[traces(stop)] = { empty^A }, I^A[traces(div)] = { empty^A , $\bot$ } ,
        I^A[traces(rec(x,call(x)))] = { empty^A, $\bot$}.

According to lemma 6.21 (3) in DPGen(TRACES) this is possible only, if there are terms t1, t2 such that:
        TRANS ⊢ trace(div) → t1,   TRANS ⊢ trace(rec(x,call(x))) → t2,
        TRANS ⊢ ↑t1,   TRANS ⊢ ↑t2
For this purpose, in TRACES the following addition may be made:
        func divergence: → Trace,

trace(div) → divergence,   trace(rec(x,call(x))) → divergence .

In appendix B another distiction between **div** and **stop** is made. Since the RAP system does not terminate if it tries to enumerate the traces of **div** and **stop**, in STEP a particular treatment for deadlock situations is introduced. The operation step there is no longer undefined when applied to a deadlock, but it delivers a special element to indicate this situation ("totalization"). This allows RAP to terminate for the process stop and other processes representing a deadlock. Divergence leads to nontermination of RAP, which corresponds well to the intuitive understanding of the processes.

7.2.3. Refusal Semantics

Within the trace semantics, internal and external nondeterminism cannot be distinguished: The traces of a system of processes remain the same, if internal nondeterminism is exchanged with external nondeterminism and vice versa. In [Olderog, Hoare 86] methods for a further refinement of the models are studied, which also can be modelled in our specification language. Below a specification is given which corresponds to the *failure model* F in [Olderog, Hoare 86] (which is the most refined model there).

A distinction between internal and external nondeterminism can be made by studying the set of actions which can be refused by a process (*refusal sets*). This means just a simple possibility to describe the behaviour of a process within various contexts (of parallel processes). For instance, the system of processes

((a → **stop**) [] (b → **stop**)) ‖{a,b} (a → **stop**)

does not lead to a deadlock under any nondeterministic choice, however in

((a → **stop**) **OR** (b → **stop**)) ‖{a,b} (a → **stop**)

a deadlock is possible (if the second alternative of **OR** is chosen). So, the process

(a → **stop**) [] (b → **stop**)

can refuse only action sets $M \subseteq (C \backslash \{a,b\})$, but the process

(a → **stop**) **OR** (b → **stop**)

can refuse all sets M where $\{a, b\} \not\subseteq M$.

This notion is made more precise by the following specification. The operation refuse nondeterministically enumerates for a given process all action sets which can be refused by the process.

spec REFUSE
basedon COM, AGENT, ID
func   refuse: Agent $\rightarrow$ **SET**(Com)
axioms
        refuse(stop) $\rightarrow$ M,
        i$\notin$M $\Rightarrow$ refuse(prefix(i,p)) $\rightarrow$ M,
        refuse(OR(p,q)) $\rightarrow$ refuse(p),
        refuse(OR(p,q)) $\rightarrow$ refuse(q),
        refuse(p) $\rightarrow$ M & refuse(q) $\rightarrow$ M $\Rightarrow$ refuse(choice(p,q)) $\rightarrow$ M,
        refuse(par(p,A,q)) $\rightarrow$ refuse(p)[A]refuse(q),
        { hiding operator omitted here! }
        refuse(rec(x,p)) $\rightarrow$ refuse(p[x/rec(x,p)])
end

Here the notation M[A]N (where M, A, N are action sets) stands for the "majority" operator introduced in [Olderog, Hoare 86]:

$$M[A]N = (M \cap A) \cup (N \cap A) \cup (M \cap N).$$

For a more detailed specification (see appendix B) a specification for sets of actions is necessary. Unfortunately such a specification cannot be described in a constructor-based style (because of the equalities which hold between the set constructors). In appendix B an implementation of sets by sequences is used, which leads to the disadvantage that a single set is represented by many different sequences.

Please note that the "hiding" operator has been omitted in REFUSE. The reason for this is that hiding cannot be easily treated within the framework of REFUSE. The combination of external nondeterminism and hiding is problematic. A close examination of the model in [Olderog, Hoare 86] shows that hiding in this context means a *non-additive* operation, and so it is basically outside the scope of our specification framework. In [Olderog, Hoare 86] a treatment can be found which is in principle transferrable to our framework, but would lead to significant overhead.

REFUSE again has essentially one model extending its primitive parts. Together with the trace semantics, it gives a description of CP, which corresponds for divergence free processes exactly to the *failure model* F in [Olderog, Hoare 86].

In a hierarchical model A$\in$DPGen(REFUSE) the following equations hold:

$\text{refuse}^A[\text{choice}(\text{prefix}(a,\text{stop}),\text{prefix}(b,\text{stop}))] =$
$\{ M \subseteq \text{Com}^A \mid a \notin M \wedge b \notin M \},$
$\text{refuse}^A\{\text{OR}(\text{prefix}(a,\text{stop}),\text{prefix}(b,\text{stop}))] =$
$\{ M \subseteq \text{Com}^A \mid a \notin M \} \cup \{ M \subseteq \text{Com}^A \mid b \notin M \},$

therefore internal and external nondeterminism are now distinguishable.

It is interesting to compare the operation refuse, when it is applied to an OR-agent and to an agent with the topmost symbol choice. In the first case, refuse delivers the nondeterministic choice between the refusal sets of both subagents, in the second case it works deterministically. This means that the construct of external nondeterminism behaves nondeterministically only in the context of trace, not in the context of refuse. Therefore the constructor choice itself should *not* be considered as truly nondeterministic. The constructor OR for internal nondeterminism, however, can be called nondeterministically by itself, since OR-terms in all contexts show exactly the same behaviour as if the following axioms were given:

$\text{OR}(p,q) \rightarrow p, \quad \text{OR}(p,q) \rightarrow q .$

The next example illustrates the (non-obvious) reason, why refusals of action sets are considered (instead of the refusal of single actions):

$\text{refuse}^A[\text{OR}(\text{OR}(\text{prefix}(a,\text{stop}),\text{prefix}(b,\text{stop})),\text{stop})] = \{ M \subseteq \text{Com}^A \},$
$\text{refuse}^A[\text{OR}(\text{prefix}(a,\text{stop}),\text{prefix}(b,\text{stop}))] =$
$\{ M \subseteq \text{Com}^A \mid a \notin M \} \cup \{ M \subseteq \text{Com}^A \mid b \notin M \}.$

An arbitrary single action can be refused by both processes, but the set of actions $\{a,b\}$ can be refused by the first process and cannot be refused by the second one.

REFUSE moreover gives a possibility to distinguish between divergent and non-divergent processes. In models $A \in \text{DPGen}(\text{REFUSE})$ the following equations hold:

$I^A[\text{refuse}(\text{stop})] = \text{Com}^A,$
$I^A[\text{refuse}(\text{div})] = \{\bot\}, \qquad I^A[\text{refuse}(\text{rec}(x,\text{call}(x)))] = \{\bot\}.$

7.3. Improvements and Applications

The example CP was intended to give some orientation with respect to the applicability of nondeterministic algebraic specifications.

First, it can be stated that a simple transition from the given description of CP to the specification is possible, and that the expressiveness of the framework is sufficient for modelling this non-trivial language. Necessary (or at least extremely useful) for this purpose were the following concepts: partial specifications, conditional axioms and hierarchical specifications.

It is also an interesting observation that all specifications within this chapter are constructor-based. So this simple sublanguage can be used successfully for non-trivial applications. The only problem with constructor-based specifications came from the question how to specify finite sets of a base set. It seems to be sensible to study the incorporation of equations between constructors.

On the other hand, it is obvious that the border of the expressiveness of our specification language is reached when some specialized semantic concepts (like the refusal semantics for the hiding-operator in CP) are considered. For semantical studies on nondeterministic programming languages, the classical mathematical framework, as it was used for CP in [Olderog, Hoare 86], always is superior. But nondeterministic algebraic specifications also bring a new aspect into the study of semantics by the new flexible and abstract approach to the *operational* aspect of nondeterministic programming languages. For the example of CP, this is illustrated by the fact that the transition semantics can be described within the same formal framework as the trace and refusal semantics.

The above-mentioned experiments are documented in the appendix B. As it was formulated as one of the aims of this work, term rewriting (or more precisely, graph rewriting, according to section 5.2) is used there as an operational semantics for the CP-specification.

Chapter 8

Concluding Remarks

In this concluding chapter, the results of this monograph are summarized and briefly evaluated. In addition to that, a number of questions are put together which could be of particular interest for future research.

8.1. Summary and Evaluation

The starting idea of this work was to integrate nondeterminism into algebraic specifications in such a way that operations in an algebra and the interpretation of terms are set-valued. Formulae do no longer denote equations but selection decisions (or inclusion relations). It has been shown that non-confluent term rewriting systems are an appropriate specification language for this world of models, if a basis of deterministic operations is designated and if the other operations work in a particularly simple (additive) way on this deterministic basis. Syntactical criteria for such additive specifications have been given. It has been shown that for this nondeterministic specification language the most important results from the theory of algebraic specifications are still valid. It has been shown, too, that the generalization can be combined with advanced concepts from algebraic specifications (for instance partiality, hierarchies) as well as from term rewriting (narrowing, graph rewriting). The relationship with logic programming has been investigated, showing that logic programming can be seen as a particular interesting subcase of the newly developed theory. For a number of examples from various areas of computer science, the basic applicability of the language has been demonstrated.

The theory presented here generalizes *equational* algebraic specifications to nodeterministic operations, leading to a kind of *inequational* specifications. This language is not such an abstract specification language that it may be used for a mathematical discussion of for instance the semantics of nondeterministic programs. The language however provides an abstract programming language in its pure form: A logical calculus is offered, which is executable by a machine (within some limitations). The algebraic specifications addressed here are *algebraic programs*.

Algebraic programs are conceptionally simple. The language is syntactically slim, the semantics is defined mathematically. Algebraic programs admit a direct approach to program verification, since a logical calculus is an elementary part of the underlying theoretical foundations. In principle, algebraic programs can be executed as fast as Prolog programs [Hanus 90]; the built-in notions of (deterministic and non-deterministic) functions enable implementations which follow very closely the mathematical semantics. These aspects designate the language which has been developed in this manuscript as the starting point for a balanced compromise between a practically usable programming language and a theoretically well-founded specification language.

8.2. Future Work

Fortunately, a number of questions, which were called "open" in earlier versions of this text [Hussmann 88/91], could be treated in detail in this book and in the recent approach [Walicki 92/93]. However, there are several possibilities to generalize the current results further, and to integrate it with other concepts for algebraic specifications. Of particular interest is here the integration of non-strict operations (at least non-strict constructors) (see for the problems arising here [Nivat 80], [Möller 82], [Broy 85], [Broy 87]). In combination with the graph rewriting techniques from section 5.2, a semantic basis for "call-by-need"-computations without a confluence condition could be obtained. Not only term reduction techniques, but also narrowing techniques should be considered for this aspect.

But also a restriction of the current approach could lead to further studies. The language of constructor-based nondeterministic specifications could serve as the basis for a powerful abstract programming language (or executable specification language). For this purpose, mainly a sufficient handling (also from the operational aspect) of constructor equations must be found. In order to achieve a good integration of nonconfluent and confluent rewriting, more general results in the spirit of lemma 5.10 (innermost normalization with a canonical subset of the axioms) are useful, for rewriting as well as for narrowing.

Another area of extension comes from the connections between the theory presented here and approaches which try to drop the distinction between sorts and objects in algebraic specifications ([Mosses 89], [Smolka 88]). Since nondeterministic operations are set-valued, they can be used for the description of sorts, too. (For instance, the operation some in example 2.20 exactly describes the sort Nat.) So polymorphic functions can be integrated into the framework of nondeterministic algebrac specifications. Such an attempt, however, will need a few technical extensions of the framework like a second kind of variables which range over sets (sort variables). This idea of a second kind of variables may also help in integrating this work more closely with the approach of [Meseguer 92].

Rewriting without confluence restrictions seems to be a framework which generalizes many important paradigms of computing. Therefore it may form a good basis for studies comparing various not obviously connected approaches to a mathematical description of computation.

References

[ADJ 78]

J. A. Goguen, J. W. Thatcher, E. G. Wagner, An initial algebra approach to the specification, correctness and implementation of abstract data types, in: R. T. Yeh (ed.), *Current trends in programming methodology, Vol. 3, Data structuring* (Prentice-Hall, Englewood Cliffs, NJ, 1978) 80-149.

[Astesiano, Costa79]

E. Astesiano, G. Costa, Sharing in nondeterminism, in: H. A. Maurer (ed.), *6th International Colloquium on Automata, Languages and Programming*, Lecture Notes in Computer Science 71 (Springer, Berlin, 1979), 1-13.

[Barbuti et al. 85]

R. Barbuti, M. Bellia, G. Levi, M. Martelli, LEAF: a language which integrates logic, equations and function, in: D. DeGroot, G. Lindstrøm (eds.), *Logic Programming: Functions, Relations and Equations* (Prentice-Hall, Englewood Cliffs, NJ, 1985) 201-238.

[Bauer, Wössner 81]

F. L. Bauer, H. Wössner, *Algorithmic Language and Program Development* (Springer, Berlin, 1981).

[Brand 75]

D. Brand, Proving theorems with the modification method, *SIAM Journal of Computing* **4** (1975) 412-430.

[Benson 79]

D. B. Benson, Parameter passing in nondeterministic recursive programs, *Journal of Computer and System Sciences* **19** (1979) 50-62.

[Bergstra, Klop 86]

J. A. Bergstra, J. W. Klop, Conditional rewrite rules: confluence and termination, *Journal of Computer and System Sciences* **32** (1986) 323-362.

[Birkhoff 35]

G. Birkhoff, On the structure of abstract algebras, *Proceedings of the Cambridge Philosophical Society* **31**(1935) 433-454.

[Bosco, Giovannetti, Moiso 88]

P. G. Bosco, E. Giovannetti, C. Moiso, Narrowing vs. SLD-Resolution, *Theoretical Computer Science* **59** (1988) 3-23.

[Broy 84]

M. Broy, Semantics of communicating processes, *Information and Control* **61** (1984) 202-246.

[Broy 85]

M. Broy, On the Herbrand-Kleene universe for nondeterministic computations, *Theoretical Computer Science* **36** (1985) 1-19.

[Broy 86]

M. Broy, A theory for nondeterminism, parallelism, communication, and concurrency, *Theoretical Computer Science* **45** (1986) 1-61.

[Broy 87]

M. Broy, Equational specification of partial higher order algebras, in: M. Broy (ed.), *Logic of programming and calculi of discrete design*, (Springer, Berlin, 1987).

[Broy, Pair, Wirsing 84]

M. Broy, C. Pair, M. Wirsing, A systematic study of models of abstract data types, *Theoretical Computer Science* **33** (1984) 139-174.

[Broy, Wirsing 81]

M. Broy, M. Wirsing, On the algebraic specification of nondeterministic programming languages, in: E. Astesiano, C. Böhm (eds.), *6th Colloquium on Trees in Algebra and Programming*, Genua 1981, Lecture Notes in Computer Science **112** (Springer, Berlin, 1981) 162-179.

[Broy, Wirsing 82]

M. Broy, M. Wirsing, Partial abstract types, *Acta Informatica* **18** (1982) 47-64.

[Cheong, Fribourg 91]

P. H. Cheong, L. Fribourg, Efficient Integration of Simplification into Prolog, in: J. Makuszy'nski, M. Wirsing (eds.), *Programming Language Implementation and Logic Programming (PLILP 91)*, Lecture Notes in Computer Science **528** (Springer, Berlin, 1991) 359-370.

[CIP 85]

The CIP Language Group, *The Munich Project CIP, Vol. I: The Wide Spectrum Language CIP-L*, Lecture Notes in Computer Science **183** (Springer, Berlin, 1985).

[Corbin, Bidoit 83]

J. Corbin, M. Bidoit, A rehabilitation of Robinson's unification algorithm, in: R. A. Mason (ed.), *Information Processing 83* (North-Holland, Amsterdam, 1983).

[Deransart 83]

P. Deransart, An operational algebraic semantics of PROLOG programs, in: *Programmation et Logique,* Proceedings (Perros-Guirrec, CNET-Lannion), 1983.

[Dershowitz 87]

N. Dershowitz, Termination of rewriting, *Journal of Symbolic Computation* **3** (1987) 69-116.

[Dijkstra 76]

E. W. Dijkstra, *A discipline of programming*, (Prentice-Hall, Englewood Cliffs, NJ, 1976).

[Fay 79]

M. Fay, First-order unification in an equational theory, in: Proceedings of the 4th Workshop on Automated Deduction, Austin, Texas, 1979.

[Floyd 67]

R. M. Floyd, Nondeterministic algorithms, *Journal of the ACM* **14** (1967) 636-644.

[Fribourg 85a]

L. Fribourg, SLOG: A logic programming language interpreter based on clausal superposition and rewriting, in: Proceedings Symposium on Logic Programming, (IEEE Computer Society Press, 1985) 172-184.

[Fribourg 85b]

L. Fribourg, A narrowing procedure for theories with constructors, in: R. E. Shostak (ed.), *7th International Conference on Automated Deduction*, Lecture Notes in Computer Science **170** (Springer, Berlin, 1985) 259-281.

[Geser 86]

A. Geser, An algebraic specification of the INTEL 8085 microprocessor: a case study, Report MIP-8608, University of Passau, Passau, 1986.

[Geser, Hussmann 86]

A. Geser, H. Hussmann, Experiences with the RAP system - A specification interpreter combining term rewriting and resolution, in: B. Robinet, R. Wilhelm (eds.): *European Symposium on Programming (ESOP) 86*, Lecture Notes in Computer Science **213** (Springer, Berlin, 1986) 339-350.

[Geser, Hussmann, Mück 88]

A. Geser, H. Hussmann, A. Mück, A compiler for a class of conditional rewrite systems, in: S. Kaplan, J.-P. Jouannaud (eds.), *Conditional Term Rewriting Systems*, Lecture Notes in Computer Science **308** (Springer, Berlin, 1988) 84-90.

[Guttag 75]

J. V. Guttag: The specification and application to programming of abstract data types, Ph. D. Thesis, University of Toronto, Report CSRG-59, Toronto, 1975.

[Hanus 90]

M. Hanus, Compiling logic programs with equality, in: P. Deransart, J. Makuszy´nski (eds.): *Programming Language Implementation and Logic Programming (PLILP 90)*, Lecture Notes in Computer Science **456** (Springer, Berlin, 1990) 387-401.

[Hansoul83]

G. E. Hansoul, A subdirect decomposition theorem for multialgebras, *Algebra Universalis*, **16** (1983) 275-281.

[Hennessy80]

M. C. B. Hennessy, The semantics of call-by-value and call-by-name in a nondeterministic environment, *SIAM Journal of Computing* **9** (1980) 67-84.

[Hennessy 90]

M. C. B. Hennessy, The semantics of programming languages (J. Wiley & Sons, Chichester, 1990).

[Hesselink88]

W. H. Hesselink, A mathematical approach to nondeterminism in data types, *ACM Transactions on Programming Languages and Systems* **10** (1988) 87-117

[Hoare 78]

C. A. R. Hoare, Communicating sequential processes, *Communications of the ACM* **21**(1978) 666-677.

[Huet, Hullot 82]

G. Huet, J.-M. Hullot, Proofs by induction in equational theories with constructors, *Journal of Computer and System Sciences* **25** (1982) 239-266.

[Huet, Oppen 80]

G. Huet, D. C. Oppen, Equations and rewrite rules: a survey, in: R. V. Book (ed.), *Formal Language Theory: Perspectives and Open Problems*, (Academic Press, New York,1980).

[Hullot 80]

J. M. Hullot, Canonical forms and unification, in: W. Bibel, R. Kowalski (eds.), *5th Conference on Automated Deduction*, Lecture Notes in Computer Science **87** (Springer, Berlin, 1980) 318-334.

[Hussmann 85/87]

H. Hussmann, *Rapid prototyping for algebraic specifications - RAP system user´s manual,* Report MIP-8504, University of Passau, Passau, 1985, 2nd extended edition 1987.

[Hussmann 88/91]

H. Hussmann, Nondeterministic algebraic specifications (in German), Ph. D. thesis, Universität Passau, 1988. English translation available as: Technical report no. TUM-I9104, Technische Universität München, 1991.

[Hussmann 92]

H. Hussmann, Nondeterministic algebraic specifications and nonconfluent term rewriting, *Journal of Logic Programming* **12** (1992) 237-255.

[Hussmann, Rank 89]

H. Hussmann, C. Rank, Specification and prototyping of a compiler for a small applicative language, in: J. A. Bergstra, M. Wirsing (eds.), *Algebraic Methods: Theory, Tools and Applications*, Lecture Notes in Computer Science **394** (Springer, Berlin, 1989) 403-418.

[Johnsson 84]

T. Johnsson, Efficient compilation of lazy evaluation, in: Proceedings of the SIGPLAN '84 Symposium on Compiler Construction, Montreal, 1984, *SIGPLAN Notices* **19** (1984), 58-69.

[Kaplan 84]

S. Kaplan, Conditional rewrite rules, *Theoretical Computer Science* **33** (1984) 175-193.

[Kaplan88]

S. Kaplan, Rewriting with a nondeterministic choice operator, *Theoretical Computer Science* **56** (1988) 37-57.

[Kapur 80]

D. Kapur, *Towards a theory for abstract data types*, Ph. D. Thesis, Massachusetts Institute of Technology, 1980.

[Knuth, Bendix 70]

D. Knuth, P. Bendix, Simple word problems in universal algebras, in: J. Leech (ed.), *Computational Problems in Abstract Algebra* (Pergamon Press, 1970) 263-297.

[Knuth, Morris, Pratt 77]

D. Knuth, J. Morris, V. Pratt, Fast pattern matching in strings, *SIAM Journal of Computing* **6** (1977) 323-350.

[Kounalis 85]

E. Kounalis, Completeness in data type specifications, in: B. Caviness (ed.), *EUROCAL '85, Proceedings Vol. 2*, Lecture Notes in Computer Science **204** (Springer, Berlin, 1985) 348-362.

[Lankford 75]

D. S. Lankford, Canonical Inference, Report ATP-32, University of Texas, 1975.

[Manna 70]

Z. Manna, The correctness of nondeterministic programs, *Artificial Intelligence* **1** (1970) 1-26.

[McCarthy 61]

J. McCarthy, A basis for a mathematical theory of computation, in: P. Braffert, D. Hirschberg (eds.), *Computer Programming and Formal Systems* (Amsterdam, 1963).

[Meseguer 92]

J. Meseguer, Conditional rewriting logic as a unified model of concurrency, *Theoretical Computer Science* **96** (1992) 73-155.

[Möller 85]

B. Möller, On the algebraic specification of infinite objects - Ordered and continuous models of algebraic types, *Acta Informatica* **22** (1985) 537-578.

[Moreno, Rodríguez 88]

J. J. Moreno-Navarro, M. Rodríguez-Artalejo, BABEL: A functional and logic language based on constructor discipline and narrowing, in: J. Grabowski, P. Lescanne, W. Wechler (eds.), *Algebraic and Logic Programming*, Lecture Notes in Computer Science **343** (Springer, Berlin, 1988), pp. 223-232.

[Mosses 89]

>P. D. Mosses, Unified algebras and institutions, in: *LICS'89, Proc.
>4th Annual Symposium on Logic in Computer Science,* IEEE (1989),
>pp. 304-312.

[Mück 90]

>A. Mück, *The compilation of narrowing,* in: P. Deransart, J.
>Makuszy´nski (eds.): *Programming Language Implementation and
>Logic Programming (PLILP 90),* Lecture Notes in Computer Science
>**456** (Springer, Berlin, 1990) 16-29.

[Narain 88]

>S. Narain, LOG(F): An optimal combination of logic programming,
>rewriting, and lazy evaluation, Internal Report, Rand Corporation,
>Santa Monica, 1988.

[Nipkow86]

>T. Nipkow, Nondeterministic data types: Models and implementations,
>*Acta Informatica* **22**(1986) 629-661.

[Nivat 80]

>M. Nivat, Nondeterministic programs: an algebraic overview, in: S. H.
>Lavington (ed.), *Information Processing 80* (North-Holland,
>Amsterdam, 1980) 17-28.

[O'Donnell 77]

>M. J. O'Donnell, *Computing in Systems Described by Equations,*
>Lecture Notes in Computer Science **58** (Springer, Berlin, 1977).

[O'Donnell 85]

>M. J. O'Donnell, *Equational logic as a programming language,* (The
>MIT Press, Cambridge, MA, 1985).

[Olderog, Hoare 86]

>E.-R. Olderog, C. A. R. Hoare, Specification-oriented semantics for
>communicating processes, *Acta Informatica* **23** (1986) 9-66.

[Padawitz 83]

>P. Padawitz, *Correctness, completeness and consistency of equational
>data type specifications,* Ph. D. Thesis, Technische Universität Berlin,
>Berlin, 1983.

[Padawitz 88]

P. Padawitz, *Computing in Horn clause theories*, EATCS Monographs in Theoretical Computer Science **16** (Springer, Berlin, 1988).

[Pickert 50]

G. Pickert, Bemerkungen zum Homomorphie-Begriff, *Mathematische Zeitschrift* **53** (1950) 375-386.

[Pickett67]

H. E. Pickett, Homomorphisms and subalgebras of multialgebras, *Pacific Journal of Mathematics* **21** (1967) 327-342.

[Pinegger 87]

T. Pinegger, *From equationally defined functions to parallel processes*, Ph. D. Thesis, Universität Passau, 1987.

[Plotkin 81]

G. D. Plotkin, A structural approach to operational semantics, Technical Report DAIMI FN-19, Computer Science Dept., Aarhus University, 1981.

[Slagle 74]

J. R. Slagle, Automated theorem proving for theories with simplifiers, commutativity and associativity, *Journal of the ACM* **21** (1974) 622-642.

[Smolka 88]

G. Smolka, *Type logic*, Lecture at the "6th Workshop on Specification of Abstract Data Types", August 1988, Berlin.

[Snyder 91]

W. Snyder, *A proof theory for general unification*, Progress in Computer Science and Applied Logic **11** (Birkhäuser, Boston, 1991).

[Subrahmanyam 81]

P. A. Subrahmanyam, Nondeterminism in abstract data types, in: S. Even, O. Kariv (eds.), *8th International Colloquium on Algorithms, Languages and Programming*, Lecture Notes in Computer Science **115** (Springer, Berlin, 1981) 148-164.

[Tamaki 84]

    H. Tamaki, Semantics of a logic programming language with a
reducibility predicate, in: *Proc. 1984 Symposium on Logic
Programming* (IEEE Computer Society Press, Washington, 1984) 259-
264.

[van Emden, Yukawa 87]

    M. H. van Emden, K. Yukawa, Logic programming with equations,
Journal of Logic Programming **4** (1987) 265-288.

[Walicki 92/93]

    M. Walicki, Calculii for nondeterministic specifications: three
completeness results, Technical Report nr. 75, Institutt for
Informatikk, Universitetet i Bergen, December 1992.

[Wirsing et al. 83]

    M. Wirsing, P. Pepper, H. Partsch, W. Dosch, M. Broy, On
hierarchies of abstract data types, *Acta Informatica* **20** (1983) 1-33.

[Wechler 91]

    W. Wechler, *Universal algebra for computer scientists*, EATCS
monographs in Theoretical Computer Science **25** (Springer, Berlin,
1991).

Appendix A: Proofs

This section contains proofs which are too long or too technical to be placed into the running text.

Proof of Lemma 1.17.1

Proof by induction on the term structure of t.

t = x, x∈X:

$$I_\beta^A[\sigma t] = \{\gamma x \mid \gamma x \in I_\beta^A[\sigma x]\} = \{e \in I_\gamma^A[t] \mid \gamma x \in I_\beta^A[\sigma x]\}.$$

t = f(t₁,...,tₙ):

$$I_\beta^A[f(\sigma t_1,...,\sigma t_n)] = \{e \in f^A(e_1,...,e_n) \mid e_i \in I_\beta^A[\sigma t_i]\} \quad \text{(definition 1.5)}$$

$$\supseteq \{e \in f^A(e_1,...,e_n) \mid e_i \in I_{\gamma_i}^A[t_i] \wedge \forall x \in \text{Vars}[t_i]: \gamma_i x \in I_\beta^A[\sigma x]\} \quad \text{(induction hyp.)}$$

$$\supseteq \{e \in f^A(e_1,...,e_n) \mid e_i \in I_\gamma^A[t_i] \wedge \forall x \in \text{Vars}[t_i]: \gamma x \in I_\beta^A[\sigma x]\} \quad (*)$$

$$= \{e \in I_\gamma^A[t] \mid \forall x \in \text{Vars}[t]: \gamma x \in I_\beta^A[\sigma x]\} \quad \text{(definition 1.5)}$$

Line (*) holds because of $\text{Vars}[t] \supseteq \text{Vars}[t_i]$.

If t is linear, t_i and t_j for $i \neq j$ have disjoint variables: $\text{Vars}[t_i] \cap \text{Vars}[t_j] = \emptyset$. Therefore the disjoint valuations γ_i can be composed to $\gamma = \gamma_1 \cup ... \cup \gamma_n$, so in line (*) the set equality holds instead of set inclusion. (Analoguous arguments apply in the induction hypothesis.) ◊

Proof of Theorem 1.19:

The proof uses a lemma which is stated below. Please note that a valuation in $W\Sigma/R$ is a substitution. Therefore, here the letter σ (where $\sigma \in SUBST(\Sigma, X)$) is used instead of β. The usual properties for substitutions are presupposed.

Lemma 1.19

(1) $\sigma t \in I_\sigma^{W\Sigma/R}[t]$

(2) $I_\sigma^{W\Sigma/R}[t] \subseteq \{ t' \in W(\Sigma, X) \mid T \vdash_{RC} \sigma t \rightarrow t' \}$

(3) $t \notin X \Rightarrow I_\sigma^{W\Sigma/R}[t] = \{ t' \in W(\Sigma, X) \mid T \vdash_{RC} \sigma t \rightarrow t' \}$

Proof of the Lemma:

Part (1):

Induction on the term structure of t

<u>$t = x, x \in X$:</u>

$\sigma t = \sigma x \in \{ \sigma x \} = I_\sigma^{W\Sigma/R}[t]$

<u>$t = f(t_1,....,t_n)$:</u>

$\sigma f(t_1,...,t_n)$

$\in \{ t' \mid \vdash f(\sigma t_1,...,\sigma t_n) \rightarrow t' \}$ (because of (REFL))

$\subseteq \{ t' \mid \vdash f(t_1',...,t_n') \rightarrow t' \wedge t_i' \in I_\sigma^{W\Sigma/R}[t_i] \}$

                                                                        (according to ind. hypothesis)

$= \{ t' \mid t' \in f^{W\Sigma/R}(t_1',...,t_n') \}$ (definition 1.15)

$= I_\sigma^{W\Sigma/R}[t]$ (definition 1.5)

Part (2):

Induction on the term structure of t

<u>$t = x, x \in X$:</u>

$I_\sigma^{W\Sigma/R}[t] = \{ \sigma x \} \subseteq \{ t' \mid \vdash \sigma x \rightarrow t' \}$ (because of (REFL))

$t = f(t_1,...,t_n)$:

$I_\sigma^{W\Sigma/R}[t] = \{t' \in f^{W\Sigma/R}(t_1',...,t_n') \mid t_i' \in I_\sigma^{W\Sigma/R}[t_i]\}$ (definition 1.5)

$\subseteq \{t' \in f^{W\Sigma/R}(t_1',...,t_n') \mid \vdash \sigma t_i \rightarrow t_i'\}$ (induction hypothesis)

$= \{t' \mid \vdash f(t_1',...,t_n') \rightarrow t' \wedge \vdash \sigma t_i \rightarrow t_i'\}$ (definition 1.18)

$\subseteq \{t' \mid \vdash f(\sigma t_1,...,\sigma t_n) \rightarrow t'\}($ (CONG), (TRANS))

$= \{t' \mid \vdash \sigma t \rightarrow t'\}.$

Part (3):

Here we have to show only the "$\supseteq$"-direction of part (2) in the case $t \notin X$. So let $t = f(t_1,...,t_n)$ and $t' \in W(\Sigma,X)$ where $\vdash \sigma t \rightarrow t'$, i.e.

$\vdash f(\sigma t_1,...,\sigma t_n) \rightarrow t'$.

So $t' \in \{t'' \mid \vdash f(\sigma t_1,...,\sigma t_n) \rightarrow t''\} \subseteq$

$\{t'' \mid \vdash f(t_1',...,t_n') \rightarrow t'' \wedge t_i' \in I_\sigma^{W\Sigma/R}[t_i]\}$

(since according to (1) $\sigma t_i \in I_\sigma^{W\Sigma/R}[t_i]$), and therefore $t' \in I_\sigma^{W\Sigma/R}[t]$.

$\Diamond$ (lemma 1.19.1)

For the proof of theorem 1.19 itself we have to show for $\triangleleft l \rightarrow r \triangleright \in R$ holds: $I_\sigma^{W\Sigma/R}[l] \supseteq I_\sigma^{W\Sigma/R}[r]$. This proof uses lemma 1.19.1:

$I_\sigma^{W\Sigma/R}[l] = \{t' \mid t' \in W(\Sigma) \wedge \vdash \sigma l \rightarrow t'\}$ (lemma 1.19.1 (3), since $l \notin X$)

$\supseteq \{t' \mid t' \in W(\Sigma) \wedge \vdash \sigma r \rightarrow t'\}$ (since $\vdash \sigma l \rightarrow \sigma r$ using (AXIOM))

$\supseteq I_\sigma^{W\Sigma/R}[r].$ (lemma 1.19.1 (2)) $\Diamond$

Proof of Theorem 2.6

Lemma 2.6.1

Let β be a valuation of X in the Σ-algebra A, σ a substitution and $Y \subseteq X$ such that

$\forall x \in Y: \qquad \mid I_\beta^A[\sigma x] \mid = 1.$

Then there is a valuation $\beta\sigma$ of X in A, defined by

$$\beta\sigma(x) = \begin{cases} I_\beta^A[\sigma x] & \text{if } x \in Y \\ \beta(x) & \text{otherwise} \end{cases}$$

and for $t \in W(\Sigma,X)$ with $\text{Vars}[t] \subseteq Y$ we have:

$$I_\beta^A[\sigma t] = I_{\beta\sigma}^A[t].$$

The lemma is proven by induction on the term structure of t (proof is omitted here).

The proof of theorem 2.6 can be performed by induction on the (length of the) derivation. The deduction rules (REFL) and (TRANS) do not pose here any problems because of the corresponding properties of set inclusion. For the other rules the following arguments can be given (identifiers are as in the definition of the deduction rule).

(CONG):

Let A be a model $A \in \text{Mod}(T)$ and $\beta \in \text{ENV}(X,A)$. The premise of the deduction rule, together with the induction hypothesis, yields:

$$I_\beta^A[t_i] \supseteq I_\beta^A[t_i'].$$

Therefore:

$$I_\beta^A[f(t_1,\dots,t_n)]$$

$$= \{e \in f^A(e_1,\dots,e_n) \mid e_j \in I^A[t_j] \text{ für } 1 \le j \le n\} \qquad\qquad \text{(def. 1.5)}$$

$$\supseteq \{e \in f^A(e_1,\dots,e_n) \mid e_j \in I^A[t_j] \text{ für } 1 \le j \le n, j \ne i, e_i \in I^A[t_i]\} \qquad \text{(premise)}$$

$$= I_\beta^A[f(t_1,\dots,t_{i-1},t_i',t_{i+1},\dots,t_n)] \qquad\qquad \text{(def. 1.5)}$$

(AXIOM-1):

Because of the premise (using the induction hypothesis):

$$\mid I_\beta^A[\sigma x] \mid = 1 \quad \text{for all } x \in \text{Vars}(l) \cup \text{Vars}(r)$$

So using lemma 2.6.1 and $A \in \text{Mod}(T)$:

$$I_\beta^A[\sigma l] = I_{\sigma\beta}^A[l] \supseteq I_{\sigma\beta}^A[r] = I_\beta^A[\sigma r].$$

(AXIOM-2):

Analoguously to (AXIOM-1) using lemma 2.6.1: $\mid I_\beta^A[\sigma t] \mid = \mid I_{\sigma\beta}^A[t] \mid = 1$.

(DET-X):

$$| I_\beta^A[x] | = | \{ \beta(x) \} | = 1.$$

(DET-D):
Because of the premise (using the induction hypothesis):

$$I_\beta^A[t1] \supseteq I_\beta^A[t2] \text{ and } | I_\beta^A[t1] | = 1,$$

So $| I_\beta^A[t2] | \le 1$. Since $| I_\beta^A[t2] | \ne 0$, we have $| I_\beta^A[t2] | = 1$.

(DET-R):
As in the case (DET-D) we have

$$I_\beta^A[t1] \supseteq I_\beta^A[t2] \text{ and } | I_\beta^A[t1] | = 1,$$

So $| I_\beta^A[t2] | = 1$, i. e. $I_\beta^A[t1] = I_\beta^A[t2]$, and therefore $I_\beta^A[t2] \supseteq I_\beta^A[t1]$.   ◊

Proof of Theorem 2.11

Lemma 2.11.1

$$\vdash f(t_1,\dots,t_n) \to t \;\wedge\; \neg(\vdash DET(f(t_1,\dots,t_n))) \;\Rightarrow$$
$$(\exists\, t_1',\dots,t_n' : \; \vdash t_i \to t_i' \;\wedge\; \vdash f(t_1',\dots,t_n') \to t \;\wedge\; \vdash DET(t_i')\,) \qquad\text{(a)}$$
$$\vee$$
$$(\exists\, t_1',\dots,t_n' : \; \vdash t_i \to t_i' \;\wedge\; \vdash f(t_1',\dots,t_n') \to t) \;\wedge\; \neg(\vdash DET(f(t_1',\dots,t_n')))$$
$$\text{(b)}$$

Proof of the Lemma:

By induction on the deduction of $\vdash f(t_1,\dots,t_n) \to t$ (the deduction rule (DET-R) can be excluded here, according to the precondition).

(REFL):
Then (b) holds trivially.

(AXIOM-1):
Then $t_i = \sigma l_i$ and $\langle f(l_1,\ldots,l_n) \rightarrow r \rangle \in R$. Because of (A1) we have $\vdash DET(l_i)$ and (since $\vdash DET(\sigma x)$ for $x \in Vars[l_i]$) also $\vdash DET(\sigma l_i)$. So (a) holds.

(CONG):
Then $t = f(t_1',\ldots,t_n')$ and $\vdash t_i \rightarrow t_i'$. If $\vdash DET(f(t_1',\ldots,t_n'))$, then because of (A2) part (a) holds. Otherwise because of $\neg(\vdash DET(f(t_1',\ldots,t_n')))$ part (b) holds.

(TRANS):
Then $\vdash f(t_1,\ldots,t_n) \rightarrow r$, $\vdash r \rightarrow t$.
The induction hypothesis can be applied to $\vdash f(t_1,\ldots,t_n) \rightarrow r$.

Case 1: (a)-part of the induction hypothesis holds.
Then using (TRANS) also the (a)-part of the claim holds.

Case 2: (b)-part of the induction hypothesis holds.
Then $r = f(r_1,\ldots,r_n)$, $\neg(\vdash DET(f(r_1,\ldots,r_n)))$, $\vdash f(r_1,\ldots,r_n) \rightarrow t$.
The induction hypothesis can be applied to $\vdash f(r_1,\ldots,r_n) \rightarrow t$ again.

Case 2.1: (a)-part of the induction hypothesis holds.
Then $\vdash r_i \rightarrow t_i'$, $\vdash DET(t_i')$, $\vdash f(t_1',\ldots,t_n') \rightarrow t$.
Using $\vdash t_i \rightarrow r_i$ (as presupposed for case 2) and (TRANS), then part (a) of the claim holds.

Case 2.2: (b)-part of the induction hypothesis holds.
Then $\vdash r_i \rightarrow t_i'$, $\neg(\vdash DET(f(t_1',\ldots,t_n')))$, $\vdash f(t_1',\ldots,t_n') \rightarrow t$. Using $\vdash t_i \rightarrow r_i$ (because of case 2) and (TRANS) then part (b) of the claim holds.

$$\lozenge \text{ (lemma 2.11.1)}$$

Theorem 2.11 follows from the lemma: Let $\vdash f(t_1,\ldots,t_n) \rightarrow t$, $\vdash DET(t)$.

Case 1: $\vdash DET(f(t_1,\ldots,t_n))$
Then because of (A-2) the condition of DET-additivity is given.

Case 2: $\neg(\vdash DET(f(t_1,\ldots,t_n)))$
Then lemma 2.11.1 can be applied. The case (b) has to be excluded (contra-

diction to |- DET(t)), so part (a) holds, which is equal to the condition of DET-additivity.   ◊

Proof of Theorem 2.14

Valuations in DΣ/R are represented by substitutions. Below the notation [σ] is used, which means ($\sigma \in$ SUBST(Σ,X)):

$\quad$ [σ]: X $\rightarrow$ W(Σ)/$\sim$ and for all x$\in$X: [σ](x) = [σx] and |- DET(σx).

Lemma 2.14.1

$\quad$ Let t$\in$W(Σ,X). Then
$$I_{[\sigma]}^{D\Sigma/R}[t] = \{ [t'] \mid t' \in W(\Sigma) \wedge |\text{-} DET(t') \wedge |\text{-} \sigma t \rightarrow t' \}$$

Proof of the Lemma:

Induction on the term structure of t:

t=x, x$\in$X:
$$I_{[\sigma]}^{D\Sigma/R}[t] = \{ [\sigma](x) \} = \{ [\sigma x] \} \subseteq \{ [t'] \mid |\text{-} DET(t') \wedge |\text{-} \sigma x \rightarrow t' \}$$

$\hfill$ (because of (REFL)).

The $\supseteq$-direction holds, since |- DET(σx) and therefore from |- σx $\rightarrow$ t' by (DET-R) follows |- t' $\rightarrow$ σx. So t'$\in$[σx].

t=f(t$_1$,...,t$_n$):
$$I_{[\sigma]}^{D\Sigma/R}[f(t_1,...,t_n)]$$

$$= \{ e \in f^{D\Sigma/R}(e_1,...,e_n) \mid e_i \in I_{[\sigma]}^{D\Sigma/R}[t_i] \} \hfill \text{(definition 1.5)}$$

$$= \{ [t'] \mid |\text{-} DET(t') \wedge |\text{-} f(t_1',...,t_n') \rightarrow t' \wedge [t_i'] \in I_{[\sigma]}^{D\Sigma/R}[t_i] \}$$

$\hfill$ (definition 2.13)

$$= \{ [t'] \mid |\text{-} DET(t') \wedge |\text{-} f(t_1',...,t_n') \rightarrow t' \wedge |\text{-} DET(t_i') \wedge |\text{-} \sigma t_i \rightarrow t_i' \}$$

$\hfill$ (induction hypothesis)

$$\subseteq \{ [t'] \mid |\text{-} DET(t') \wedge |\text{-} \sigma f(t_1,...,t_n) \rightarrow t' \} \hfill \text{(by (CONG),(TRANS))}$$

For the-direction, let $t = f(t_1, \ldots, t_n)$ and $t' \in W(\Sigma)$ such that $\vdash \text{DET}(t')$ and $\vdash \sigma f(t_1, \ldots, t_n) \to t'$.

Because of the DET-additivity there exist $t_1', \ldots, t_n'$ such that

$\vdash \text{DET}(t_i'), \vdash \sigma t_i \to t_i', \vdash f(t_1', \ldots, t_n') \to t$, i. e. $t' \in I_{[\sigma]}^{D\Sigma/R}[t]$.

$$\Diamond \text{ (lemma 2.14.1)}$$

For the proof of theorem 2.14, let $\langle \text{DET}(t1) \rangle \in R$, $[\sigma]$ a valuation. Then:

$\vert I_{[\sigma]}^{D\Sigma/R}[t1] \vert = \vert \{ [t'] \mid \vdash \text{DET}(t') \wedge \vdash \sigma t1 \to t' \} \vert$ (lemma 2.14.1)

$= \vert \{ [t1] \} \vert$ (because of $\vdash \text{DET}(\sigma t1)$, rule (DET-R))

$= 1.$

Let $\langle t1 \to t2 \rangle \in R$, $[\sigma]$ a valuation. Then:

$I_{[\sigma]}^{D\Sigma/R}[t1] = \{ [t'] \mid \vdash \text{DET}(t') \wedge \vdash \sigma t1 \to t' \}$ (lemma 2.14.1)

$\supseteq \{ [t'] \mid \vdash \text{DET}(t') \wedge \vdash \sigma t2 \to t' \}$ (since $\vdash \sigma t1 \to \sigma t2$ by (AXIOM-1))

$= I_{[\sigma]}^{D\Sigma/R}[t2]$ (lemma 2.14.1).

$D\Sigma/R$ is term-generated: Let $[t] \in s^{D\Sigma/R}$. Then $\vdash \text{DET}(t)$ and therefore:

$I^{D\Sigma/R}[t] = \{ [t'] \mid \vdash \text{DET}(t') \wedge \vdash t \to t' \}$ (lemma 2.14.1)

$= \{ [t] \}$ (rules (REFL), (DET-R)).$\Diamond$

Proof of Theorem 3.13

φ is a loose Σ-homomorphism:

$\{ e \in \varphi([t]) \mid [t] \in f^{D\Sigma/R}([t_1], \ldots, [t_n]) \}$

$= \{ e \in I^A[t] \mid \vdash f(t_1, \ldots, t_n) \to t \wedge \vdash \text{DET}(t)$ (definition of φ, definition 2.13)

$\subseteq I^A[f(t_1, \ldots, t_n)]$ (theorem 2.6: $A \models f(t_1, \ldots, t_n) \to t$)

$= \{ e \in f^A(e_1, \ldots, e_n) \mid e_i \in I^A[t_i] \}$ (definition 1.5)

$= \{ e \in f^A(e_1, \ldots, e_n) \mid e_i \in \varphi[t_i] \}$ (definition of φ).

φ is unique: Let ψ be another homomorphism $\psi: D\Sigma/R \to A$, then we have:

$\psi([t]) \subseteq I^A[t]$ for all $t \in s^{D\Sigma/R}$.   (*)

Proof of the line (*) by induction on the term structure of t:

Let $t = f(t_1,\ldots,t_n)$. Since $[t] \in s^{D\Sigma/R}$, $\vdash DET(t)$ holds. Using rule (REFL) this means

$$[t] \in \{[t'] \mid t' \in W(\Sigma) \wedge \vdash DET(t') \wedge \vdash t \to t'\}.$$

Therefore

$$\psi([t]) \subseteq \{e \in \psi([t']) \mid t' \in W(\Sigma) \wedge \vdash DET(t') \wedge \vdash t \to t'\}$$
$$= \{e \in \psi([t']) \mid t' \in W(\Sigma) \wedge \vdash DET(t') \wedge \vdash f(t_1,\ldots,t_n) \to t'\}$$
$$= \{e \in \psi([t']) \mid [t'] \in f^{D\Sigma/R}([t_1],\ldots,[t_n])\} \qquad \text{(definition 2.13)}$$
$$\subseteq \{e \in f^A(e_1,\ldots,e_n) \mid e_i \in \psi([t_i])\} \qquad \text{(since } \psi \text{ is homomorphism)}$$
$$\subseteq \{e \in f^A(e_1,\ldots,e_n) \mid e_i \in I^A[t_i]\} \qquad \text{(induction hypothesis)}$$
$$= I^A[f(t_1,\ldots,t_n)] \qquad \text{(def. 1.5)}.$$

Since $\vdash DET(t)$, from theorem 2.6 follows $|I^A[t]| = 1$. Since $\psi([t]) \neq \emptyset$, from (*) follows:

$$\psi([t]) = I^A[t] = \varphi([t]) \qquad \text{for all } t \in s^{D\Sigma/R}. \qquad \Diamond$$

Proof of Lemma 3.18

(1) ⇒ (2):
Let A be maximally deterministic, $B \in Gen(T)$, $\varphi: B \to A$ a loose homomorphism. Then for a $t \in W(\Sigma)$:

$$|\{e' \in \varphi(e) \mid e \in I^B[t]\}| \geq |I^B[t]| \geq |I^A[t]|$$

(since A is maximally deterministic) and

$$\{e' \in \varphi(e) \mid e \in I^B[t]\} \subseteq I^A[t]$$

(since φ loose homomorphism) So φ is a tight homomorphism.

(2) ⇒ (3):
Since T is DET-additive and DET-complete, $D\Sigma/R \in Gen(T)$. I^A is a loose homomorphism from $D\Sigma/R$ to A.

Due to the precondition, I^A is a tight homomorphism. With lemma 2.14.1:

$$\{e \in I^A[t'] \mid \vdash t \to t' \wedge \vdash DET(t')\} = I^A[t].$$

So for an $e \in I^A[t]$ there is a term t' such that

$$\vdash t \to t' \wedge \vdash DET(t') \text{ and } e \in I^A[t'], \text{ i. e. } I^A[t'] = \{e\}.$$

(3) ⇒ (1):
Let $B \in Gen(T)$, $\varphi: B \to A$ loose homomorphism, $t \in W(\Sigma)$.

According to the precondition, for $e \in I^A[t]$ there is a term t' such that

$$I^A[t'] = \{ e \} \wedge \vdash t \to t' \wedge \vdash DET(t').$$

So there is a e'$\in s^B$ such that

$$I^B[t'] = \{ e' \} \wedge e' \in I^B[t] .$$

The condition for a homomorphism yields $\varphi(e') \subseteq \{ e \}$, i. e. $\varphi(e') = \{ e \}$. So there is a surjective pointwise mapping from $I^B[t]$ to $I^A[t]$, and therefore

$$| I^A[t] | \leq | I^B[t] | . \qquad \Diamond$$

Proof of Theorem 4.8

The proof is performed in exact analogy to the proof of theorem 2.6, except of the newly added rule. Using the identifiers from the proof of theorem 2.6 and from definition 4.7 we have:

(AXIOM-1-COND):

Let $V =_{def} Vars(l) \cup Vars(r) \cup Vars(t_1) \cup ... \cup Vars(t_n) \cup Vars(t_1') \cup ... \cup Vars(t_n'))$.
Because of the premise (using the induction hypothesis)

$$| I_\beta^A[\sigma x] | = 1 \quad \text{for all } x \in V$$

Because of the other premises and lemma 2.6.1

$$I_{\sigma\beta}^A[t_i] = I_\beta^A[\sigma t_i] \supseteq I_\beta^A[\sigma t_i'] = I_{\sigma\beta}^A[t_i'] \qquad \text{for } i \in \{1,...,n\}$$

Since $A \in Mod(T)$:

$$I_\beta^A[\sigma l] = I_{\sigma\beta}^A[l] \supseteq I_{\sigma\beta}^A[r] = I_\beta^A[\sigma r]. \qquad \Diamond$$

Proof of Theorem 4.11:

The construction of a term model is exactly analoguous to theorem 2.14, using a similar lemma. The only difference appears during the proof for the validity of conditional axioms:

Let $\langle t_1 \to t_1' \& ... \& t_n \to t_n' \Rightarrow l \to r \rangle \in R$, $[\sigma]$ a valuation, which fulfills the conditions, i. e. for $1 \leq i \leq n$:

$$I_{[\sigma]}^{D\Sigma/R}[t_i] \supseteq I_{[\sigma]}^{D\Sigma/R}[t_i'] .$$

Because of ⊢-COND DET(t_i') (simplicity condition) and ⊢-COND DET(σx) for $x \in X$ (due to the construction of DΣ/R) we have ⊢-COND DET(σt_i'). (This needs a simple lemma, which can be proven by induction on the term structure.). So $\sigma t_i' \in I_{[\sigma]}^{D\Sigma/R}[t_i'] \subseteq I_{[\sigma]}^{D\Sigma/R}[t_i]$, i. e.:

$$\text{⊢-COND } \sigma t_i \to \sigma t_i' .$$

Therefore (AXIOM-1-COND) can be applied, and gives ⊢-COND $\sigma l \to \sigma r$, and so
$$I_{[\sigma]}^{D\Sigma/R}[l] \supseteq I_{[\sigma]}^{D\Sigma/R}[r].$$

The proof of
$$\text{DΣ/R} \models t1 \to t2 \quad \Leftrightarrow \quad \text{DGen(T)} \models t1 \to t2$$
is completely analoguous to theorems 2.23 / 2.27.   ◊

Proof of Lemma 4.26

By induction on the length of the derivation for ⊢- $\sigma t1 \to t2$:

(REFL):
In this case, $t2 = \sigma\, t1$. Choose $\sigma' = \iota$, $\lambda = \sigma$, $t2' = t1$, $V' = V$. Then by (REFL-N) ⊢- $t1 -N\to_\iota t1$ and

(i)   Vars[t2']=Vars[t1] $\subseteq$ V',
         Dom[λ] = Dom[σ] $\subseteq$ V' (by given preconditions),
         Vars[σ'] = Ø $\subseteq$ V',
(ii) $\sigma =_{[V]} \sigma\iota = \lambda\,\sigma$' (trivially),
(iii) $t2 = \sigma\, t1 = \lambda\, t2$' (due to the special case).

(TRANS):
In this case, there is a t3 such that
$$\text{⊢- }\sigma t1 \to t3, \qquad \text{⊢- } t3 \to t2.$$
Using the induction hypothesis on the first one of these derivations, we have $\lambda 1, \sigma 1 \in \text{SUBST}(\Sigma_C, X)$, $t3' \in W(\Sigma, X)$, $V1 \supseteq V$ such that
$$\text{⊢- } t1 -N\to_{\sigma 1} t3', \text{Vars[t3']} \subseteq V1, \text{Dom[}\lambda 1] \subseteq V1, \text{Vars[}\sigma 1] \subseteq V1,$$
$$\sigma =_{[V]} \lambda 1\,\sigma 1, t3 = \lambda 1\, t3'.$$
This admits the application of the induction hypothesis to the second derivation (for ⊢- $\lambda 1$ t3' → t2), giving: $\lambda, \sigma 2 \in \text{SUBST}(\Sigma_C, X)$, $t2' \in W(\Sigma, X)$, $V' \supseteq V1 \supseteq V$ such that

$\vdash t3' -N\rightarrow_{\sigma 2} t2'$, $\text{Vars}[t2']\subseteq V'$, $\text{Dom}[\lambda]\subseteq V'$, $\text{Vars}[\sigma 2]\subseteq V'$,

$\lambda 1 = {}_{[V1]} \lambda \sigma 2$, $t2 = \lambda t2'$.

Now define $\sigma' = \sigma 2 \sigma 1$, then by (TRANS-N) $\vdash t1 -N\rightarrow_{\sigma'} t2'$.

(i)   The first two parts are already given by the second induction step. For
        the third part of (i), $\text{Vars}[\sigma']=\text{Vars}[\sigma 2 \sigma 1]\subseteq \text{Vars}[\sigma 1]\cup\text{Vars}[\sigma 2] \subseteq V'$.

(ii)   We have $\sigma ={}_{[V]} \lambda 1 \sigma 1$ and $\lambda 1 = {}_{[V1]} \lambda \sigma 2$. Since $\text{Dom}[\lambda 1]\subseteq V1$, this
        means $\lambda 1 = \lambda \sigma 2$, hence $\sigma ={}_{[V]} \lambda \sigma 2 \sigma 1 = \lambda \sigma'$.

(iii)   Given by the second induction step.

(CONG):

For the sake of simplicity, we assume without loss of generality the typical case
of $n = 2$, $i = 2$, $t1 = f(u_1,u_2)$.

So there is an r_2 such that

$\sigma t1 = f(\sigma u_1,\sigma u_2)$, $t2 = f(\sigma u_1,r2)$ and $\vdash \sigma u_2 \rightarrow r_2$.

By induction hypothesis there are $\lambda, \sigma'\in\text{SUBST}(\Sigma_C, X)$, r_2', V' with

$\vdash u_2 -N\rightarrow_{\sigma'} r_2'$, $\text{Vars}[r_2']\subseteq V'$, $\text{Dom}[\lambda]\subseteq V'$, $\text{Vars}[\sigma']\subseteq V'$,

$\sigma ={}_{[V]} \lambda\sigma'$, $r_2 = \lambda r_2'$.

Choose now $t2' = f(\sigma'u_1,r_2')$. By (CONG-N) holds

$\vdash t1 = f(u_1,u_2) -N\rightarrow_{\sigma'} f(\sigma'u_1,r_2') = t2'$.

(i) $\text{Vars}[t2']\subseteq\text{Vars}[\sigma']\cup\text{Vars}[u_1]\cup\text{Vars}[r_2']\subseteq V'$, due to the induction step
        and $\text{Vars}[u_1]\subseteq\text{Vars}[t1]\subseteq V\subseteq V'$. The other parts are given by the
        induction step.

(ii)   Given by the induction step.

(iii) $\lambda t2' = f(\lambda\sigma'u_1, \lambda r_2') = f(\sigma u_1,r_2) = t2$, due to the induction step and
        $\text{Vars}[u_1]\subseteq V$.

(AXIOM):

This means that there is an axiom $\triangleleft l \rightarrow r \triangleright \in R$ and a constructor substitution
$\tau\in\text{SUBST}(\Sigma_C,X)$ such that $\sigma t1 = \tau l$, $t2 = \tau r$. It follows immediately that $t1\notin X$.
(If $t1 = x$, $x\in X$, then σx contains at least the topmost symbol of l which is a
non-constructor in a constructor-based system. This contradicts to
$\sigma\in\text{SUBST}(\Sigma_C,X)$).

The substitution τ can always be splitted into another substitution τ' and a
renaming ρ (which assigns "fresh" names to all the variables of the axiom) such
that

$\tau'\rho = \tau$, $\text{Dom}(\tau')\cap V = \emptyset$, $\text{Dom}(\sigma)\cap(\text{Vars}(\rho l)\cup\text{Vars}(\rho r)) = \emptyset$, and
$\text{Dom}(\tau')\cap\text{Dom}(\sigma) = \emptyset$.

Then $\tau' \cup \sigma$ is well-defined and

$$(\tau' \cup \sigma)t1 = \sigma t1 = \tau l = \tau' \rho l = (\tau' \cup \sigma)(\rho l).$$

Thus t1 and ρl are unifiable by the unifier $(\tau' \cup \sigma)$ (which is in $\text{SUBST}(\Sigma_C, X)$, since τ and σ are in there). Define σ' as the most general unifier of t1 and ρl, and η as the corresponding specialization, such that:

$$\eta \sigma' = (\tau' \cup \sigma).$$

Obviously, η and σ' are in $\text{SUBST}(\Sigma_C, X)$ (as $\tau' \cup \sigma$ is). Rule (AXIOM-N) yields now $\vdash t1 \ _{N}\!\!\rightarrow_{\sigma'} t2'$. Define the remaining items as

$$\lambda = \eta|_{V'}, \ t2' = \sigma' \rho r, \ V' = V \cup \text{Vars}[\sigma'] \cup \text{Vars}[\rho].$$

(i)   Since ρ renames all variables of the axiom, $\text{Vars}[t2'] = \text{Vars}[\sigma' \rho r] \subseteq \text{Vars}[\sigma'] \cup \text{Vars}[\rho] \subseteq V'$. By definition, we have $\text{Dom}[\lambda] = \text{Dom}[\eta|_{V'}] \subseteq V'$ and $\text{Vars}[\sigma'] \subseteq V'$.

(ii)   Since $\text{Dom}(\tau') \cap V = \varnothing$, $\sigma =_{[V]} (\tau' \cup \sigma) = \eta \sigma'$. Since $\text{Vars}[\sigma'] \subseteq V'$, $\eta \sigma' = (\eta|_{V'}) \sigma' = \lambda \sigma'$.

(iii) $\lambda t2' = (\eta|_{V'}) \sigma' \rho r = \eta \sigma' \rho r = (\tau' \cup \sigma) \rho r = \tau' \rho r = \tau r = t2.$ $\Diamond$

Proof of Lemma 4.39:

The following proof omits technical detail at a few points. A more general proof can be found in the literature ([Bosco et al. 88]). The proof of the lemma proceeds by induction on the derivation in the narrowing calculus.

(REFL-N):

In this case, t1 = t2 and $\theta = \iota$. It is sufficient to choose $\theta' = \theta = \iota$.

(TRANS-N):

Here we have

$$\vdash t1 \ _{N}\!\!\rightarrow_{\theta 1} t2', \ \vdash t2' \ _{N}\!\!\rightarrow_{\theta 2} t2 \text{ and } \theta = \theta 2 \ \theta 1.$$

Let $\Phi[t2'] = (c2', B2')$. By induction hypothesis

B1 **where** $\sigma \vdash$ B2' **where** $\theta 1' \sigma$ and

B2' **where** $\sigma' \vdash$ B2 **where** $\theta 2' \sigma'$.

This means by transitivity of resolution steps

B1 **where** $\sigma \vdash$ B2 **where** $\theta' \sigma$,

if θ' is defined as $\theta' = \theta 2' \ \theta 1'$. Using the other parts of the induction hypothesis and the variable restrictions for $\lambda 1, \lambda 2$:

$$\theta' c1 = \theta 2' \theta 1' c1 = \theta 2' c2' = c2;$$

$$\theta' = \theta 2' \ \theta 1' = (\theta 2 \cup \lambda 2)(\theta 1 \cup \lambda 1) = \theta 2 \ \theta 1 \cup \lambda 2 \ \lambda 1 = \theta \cup \lambda 2 \ \lambda 1.$$

(CONG-N):

Using a typical subcase ($n=2$, $i=2$) , we have here

$t1 = f(u1, u2)$, $t2 = f(\theta u1, u2')$, and $u2 \xrightarrow{N}_\theta u2'$.

Let $\Phi[u1] = (d1, C1)$, $\Phi[u2] = (d2, C2)$, $\Phi[u2'] = (d2', C2')$. By induction hypothesis,

C2 **where** σ |- C2' **where** $\theta'\sigma$

and $\theta'd2 = d2'$.

Case 1: $f \in C$

In this case,

$\Phi[t1] = (\ f(d1,d2), C1 \bullet C2\ )$, $\Phi[t2] = (\ f(\theta d1,d2'), \theta C1 \bullet C2'\ )$

(using the fact that $\Phi[t] = (c,B) \Rightarrow \Phi(\theta t) = (\theta c, \theta B)$, and abstracting from renamings into "fresh" variables). This means $B1 = C1 \bullet C2$, $B2 = \theta C1 \bullet C2'$. Using the induction hypothesis and (RES),

B1 **where** σ |- $\theta'C1 \bullet C2'$ **where** $\theta'\sigma$.

The variable restrictions for λ ensure that $\theta'C1 = \theta C1$, so

B1 **where** σ |- B2 **where** $\theta'\sigma$.

Moreover $\theta'c1 = \theta'f(d1,d2) = f(\theta'd1,\theta'd2) = f(\theta'd1,d2') = c2$.

Case 2: $f \notin C$

In this case,

$\Phi[t1] = (z, f(d1,d2,z) \bullet C1 \bullet C2)$, $\Phi[t2] = (z, f(\theta d1,d2',z) \bullet \theta C1 \bullet C2')$

(abstracting from renamings into "fresh" variables again). This means $B1 = f(d1,d2,z) \bullet C1 \bullet C2$, $B2 = f(\theta d1,d2',z) \bullet \theta C1 \bullet C2'$. Using the induction hypothesis,

B1 **where** σ |- $f(\theta'd1,d2',\theta'z) \bullet \theta'C1 \bullet C2'$ **where** $\theta'\sigma$.

Since z is a fresh variable, $\theta'z=z$. The variable restrictions for λ ensure that $\theta'd1 = \theta d1$ and $\theta'C1 = \theta C1$, so

B1 **where** σ |- B2 **where** $\theta'\sigma$.

Moreover, $\theta'c1 = \theta'z = \theta z = z = c2$.

(AXIOM-N):

In this case,

$t1 = f(d_1,...,d_n)$, $t2 = \theta r$ and $\langle f(c_1,...,c_n) \rightarrow r \rangle \in R$,

θ is a mgu. of t1 and $f(c_1,...,c_n)$ (omitting the technicalities of the renaming ρ). Since $f \notin C$ and the d_i are constructor terms, $\Phi[t1] = (\ z, f(d_1,...,d_n,z)\ )$, i.e. $B1 = f(d_1,...,d_n,z)$, $c1 = z$. Due to the construction of $\Phi(T)$, there is a program clause $\langle f(c_1,...,c_n,c) :- B \rangle$ where $(c, B) = \Phi[r]$. Define now $\theta' = [c / z]\theta$, which is a mgu. of $f(c_1,...,c_n,c)$ and $f(d_1,...,d_n,z)$. Using (RES),

B1 **where** σ |- $\theta'B$ **where** $\theta'\sigma$.

We have $\Phi[t2] = \Phi[\theta r] = (\theta c, \theta B)$, i.e. $B2 = \theta B = \theta'B$, which shows that the derivation from above is the needed one. Moreover, $c2 = \theta c = c$, due to the fact that c contains only variables created during the flattening of r, which are not affected by θ. So $\theta'c1 = \theta'z = c = c2$.   ◊

Proof of Lemma 5.10:

Part (1):

First we show the following proposition:
$$\forall u \in Occ[t], \langle l \to r \rangle \in R\backslash D, \sigma \in SUBST(\Sigma_C): \tag{1}$$
$$t1/u = \sigma l \Rightarrow (\downarrow_D^{im}[t1])/u = \sigma l .$$

The proof is done by induction on the length of an arbitrary term rewriting sequence, which reduces t1 to (D-innermost-)normal form:
$$t1 = t_0 \to_D^{im} t_1 \to_D^{im} t_2 \to_D^{im} \cdots \to_D^{im} t_n = \downarrow_D^{im}[t1] .$$

So we have to show (by induction on n):
$$\forall u \in Occ[t], \langle l \to r \rangle \in R\backslash D, \sigma \in SUBST(\Sigma_C): t1/u = \sigma l \Rightarrow t_n/u = \sigma l .$$

n = 0:
Here $t_n/u = t_0/u = t1/u = \sigma l$, according to the precondition.

n > 0:
In this case, there are $v \in Occ[t_{n-1}], \langle l_D \to r_D \rangle \in D, \sigma_D \in SUBST(\Sigma_C)$ such that
$$t_{n-1}/v = \sigma_D l_D, t_n = t_{n-1} [v \leftarrow \sigma r_D].$$
Let $t1/u = \sigma l$, then the induction hypothesis yields $t_{n-1}/u = \sigma l$. The nonoverlapping condition allows to exclude the case $u = v$. Due to the innermost rewriting $(\sigma, \sigma_D \in SUBST(\Sigma_C))$ it is impossible that u is a prefix of v or reversely. So u and v are independent occurrences, and we have:
$$t_n/u = (t_{n-1}[v \leftarrow \sigma_D r_D])/u = \sigma l .$$
Due to the precondition there are fixed $u \in Occ[t], \langle l \to r \rangle \in R\backslash D, \sigma \in SUBST(\Sigma_C)$ such that $t1/u = \sigma l$.

Part (2):

Now we can show by an analoguous induction:
$$(\downarrow_D^{im}[t1])[u \leftarrow \sigma r] \to_D^{im*} \downarrow_D^{im}[t1[u \leftarrow \sigma r]] \tag{2}$$

n = 0:

Here $t_n[u \leftarrow \sigma r] = t_0[u \leftarrow \sigma r] = t1[u \leftarrow \sigma r] \xrightarrow{im*}_D \downarrow^{im}_D [t1[u \leftarrow \sigma r]]$

(by the definition of $\downarrow^{im}_D$).

n > 0:

There are $v \in Occ[t_{n-1}]$, $\langle l_D \rightarrow r_D \rangle \in D$, $\sigma_D \in SUBST(\Sigma_C)$ such that
$t_{n-1}/v = \sigma_D l_D$, $t_n = t_{n-1}[v \leftarrow \sigma r_D]$.
From (1) follows that $t_n/u = \sigma l$. As above, v and u are independent. So, according to the induction hypothesis:

$$t_n[u \leftarrow \sigma r] \xrightarrow{im}_D t_n[v \leftarrow \sigma_D r_D, u \leftarrow \sigma r]$$

$$= t_{n-1}[u \leftarrow \sigma r] \xrightarrow{im*}_D \downarrow^{im}_D [t1[u \leftarrow \sigma r]].$$

If we choose t2' $= (\downarrow^{im}_D [t1])[u \leftarrow \sigma r]$, so line (1) gives the fact

$\downarrow^{im}_D [t1] \xrightarrow{im}_{R \backslash D} t2'$. Line (2) means here that t2' $\xrightarrow{im*}_D \downarrow^{im}_D [t2]$,

and because of the normal form property of $\downarrow^{im}_D$ follows:

$$\downarrow^{im}_D [t2'] = \downarrow^{im}_D [t2]. \qquad\qquad\qquad\qquad \Diamond$$

Proof of Theorem 5.17:

The proof sketch uses the identifiers from definition 5.15. Let val3 be the given valuation for the graph G_3. In a first step, a valuation val' for the graph G is constructed, together with an environment β', assigning values to the variables in the axiom (Vars[l]∪Vars[r]). Let n∈N. We start with a valuation val, assigning values only to the nodes in $Glroot_R$.

lab(n)∈X:

This means that x∈Vars[l]∪Vars[r], $n = n_x$. Due to the variable-restriction x∈Vars[l], therefore $\phi(n_x)$ is defined.

x∈Vars[r]:

Then $\phi(n_x)$ is reachable from $root_2$, due to the construction of G_2. Hence $val3(\phi(n_x))$ is defined. In this case $val(n_x) =_{def} val3(\phi(n_x))$.

x∉Vars[r]:

Then the variable x "deletes" some term at axiom application. Let val''
be an arbitrary extension of val_3 to the nodes reachable from the ϕ-
images of these variables, and $val(n_x) =_{def} val''(\phi(n_x))$.

lab(n)∉X:

·   Since we treat here only nodes reachable from $root_R$, n is reachable
from $root_2$, and $val(n) =_{def} val_3(n)$.

From val, a definition of the valuation β' can be derived, by $β'(x) =_{def} val(n_x)$.
Using this environment, $val(root_R) \in I^A_{β'}[l]$, and therefore $val(root_R) \in I^A_{β'}[r]$.

So there is an extension val' of val to all variables in G (including those not
reachable from $root_R$) such that $val'(root_L) = val'(root_R)$. It is an important
consequence of the constructor-based form of the axiom that the valuation val' is
uniquely determined for all nodes in the left hand side, except of $root_L$. More
formally:

$$\forall n \in N: \text{ n reachable from } root_L \text{ and } n \neq root_L \Rightarrow$$
$$val'(n) = (val_3 \cup val'')(\phi(n))$$

(Exact proof by induction on the structure of the graph representing the right
hand side.)

Using these preliminaries, the given valuation val_3 can be extended to a
valuation val_2 for G_2. For this purpose, we have to study the nodes $n \in N_2 \backslash N_3$
removed by garbage collection. There are two cases:

Case 1: Parts of G_0 "deleted" by the rule application:
In this case, there is an $x \in Vars[l] \backslash Vars[r]$, and n is reachable from $\phi(n_x)$, there is
no other way to reach n from $root_2$. Here $val_2(n) =_{def} val''(n)$ (as above).

Case 2: Parts of the pattern which are replaced:
In this case, there is an $n' \in N$, such that n is reachable from $root_L$ and $\phi(n') = n$.
Here $val_2(n) =_{def} val'(\phi^{-1}(n))$. This definition is admissible only if $val'(\phi^{-1}(n))$
is unique. For $n' \neq root_L$, this is ensured by the above-mentioned observation
(which relies on the constructor-based form). For $root_L$ the definition is unique,
since $\phi^{-1}(root_L)$ is unique.

We have to check now that the constructed val_2 is a proper valuation of G_2. For
variable nodes, this is obvious from $val_2(n_x) = val_3(n_x) = βx$. For a non-

variable node n (i.e. lab(n) = f), the only problematic case is where a node $n \in N$ has a successor $arg(n)_i \in N_0$. This is only the case if there is a $n_x \in N$ such that $arg(n)_i = \phi(n_x)$. Therefore $val_2(arg(n)_i) = val(x) = val'(x) = \beta'x$, which means that val_2 takes its value for $arg(n)_i$ from a proper valuation for G. For the nodes "below" $arg(n)_i$, val_2 takes its values from a proper valuation of G_3 (if $x \in Vars[r]$) or $G_2|\phi(n_x)$ (if $x \notin Vars[r]$).

In order to validate the step from a valuation val_2 for G_2 to a valuation val_1 for G_1, we simply define $val_1 =_{def} val_2$. The valuation val_1 is also a proper valuation for G_1, since the values assigned to $\phi(root_L)$ (the node to be replaced) and n_r (the replacement node) are equal:

$$val_2(\phi(root_L)) = val'(root_L) = val(root_R).$$

Here we have two cases:

Case 1: $lab(root_R) \in X$:
Then $val(root_R) = val_3(\phi(root_R)) = val_3(n_r) = val_2(n_r).$

Case 2: $lab(root_R) \notin X$:
Then $val(root_R) = val_3(root_R) = val_3(n_r) = val_2(n_r).$

The final step from val_1 to a valuation val_0 for G_0 is achieved by $val_0 =_{def} val_1|N_0$.

It remains to show that $val_0(root_0) = val_3(root_3)$. Again there are two cases:

Case 1: Rewriting below the root of G_0:
This means that $root_3 \in N_0$ and $root_0 = root_3$. Then $val_3(root_3) = val_3(root_0) = val_0(root_0)$ (since val_3 coincides with val_0 on N_0).

Case 2: Rewriting at the root of G_0:
This means that $root_3 = n_r$ and $root_0 = \phi(root_L)$. Then $val_3(root_3) = val_3(n_r) = val_2(\phi(root_L)) = val_2(root_0) = val_0(root_0)$.   ◊

Proof of Theorem 5.21:

It suffices here to show the lemma
$$wf[G1] \wedge TM[G1] = t1 \wedge \vdash_C t1 \rightarrow t2 \Rightarrow$$

$$\exists \ G2: wf[G2] \wedge TM[G2] = t2 \wedge G1 \xrightarrow[GR]{*} G2.$$

The proof for this lemma proceeds by induction on the $\vdash_C$-derivation for the formula $t1 \rightarrow t2$. The case (REFL) is trivial, (TRANS) is obvious by the syntactical form of the lemma. The remaining cases are (CONG) and (AXIOM-1-C), where the (CONG) case must show that the anomaly from example 5.18 cannot happen under the given preconditions.

(CONG):
In this case, $t1 = f(t_1,\ldots,t_i,\ldots,t_n)$, $t2 = f(t_1,\ldots,t_i',\ldots,t_n)$. G1 contains a subgraph $G_i = G1|arg(root1)_i$ such that $TM[G_i] = t_i$. By induction hypothesis, there is a graph G_i' such that $G_i \xrightarrow[GR]{*} G_i'$. There are two cases (let $root_i$ be the root of G_i):

Case 1: is_shared(r_i):
Since wf[G1] (and therefore wf[G_i]), G_i is irreducible with constructor-based axioms. Hence $G_i' = G_I$, and the claim of the lemma holds trivially.

Case 2: $\neg$is_shared(r_i):
In this case, none of the t_j ($j \neq i$) is represented by the same graph as t_i. Therefore the same graph context building G1 from G_i can be built around G_i', giving G2. Since there is only one path from the root of G2 to the replaced parts (in G_i'), $TM[G2] = t2$ and wf[G2].

(AXIOM-1-C)
In this case, lab(root1) = f, $f \notin C$ (due to the constructor-based form of the axioms), and $\lhd \rightarrow r \in R$ such that $t1 = \sigma l$. A redex for $GR(\lhd \rightarrow r)$ in G1 is constructed by $\phi(l_u) = G1/u$, and by $\phi(n_x) = G1/u_x$, where u_x is the unique position of the variable x in l (due to left-linearity). Here we use the convention

    $G/\varepsilon =_{def}$ root,

    $G/i{\bullet}u =_{def} (G|arg(root)_i)/u)$.

The graph G2 is constructed due to definition 5.15. Since σ is a constructor-substitution, $G1|u_x$ contains only constructor- and variable-labels. Together with the construction of $GR(\lhd \rightarrow r)$ this means wf[G2]. The fact that $TM[G2] = \sigma r = t2$ is obvious from the construction of $GR(\lhd \rightarrow r)$. $\Diamond$

Proof of Theorem 6.6:

The lemma 2.6.1 can be taken over from the proof of theorem 2.6, if for the substitution σ we claim additionally:

$$\bot \notin I_\beta^A[\sigma x] \;\wedge\; |\, I_\beta^A[\sigma x]\, | = 1.$$

The premises in the deduction rules (AXIOM-1-D) and (AXIOM-2-D) allow the applicability of the lemma, even after this extension. Except of this, the proof of theorem 2.6 can be identically adapted for the rules which belong to the "total" calculus. For the new rules we have:

(AXIOM-3-D):
Using the analoguous lemma to lemma 2.6.1, with $A \in \mathrm{Mod}(T)$:

$$\bot \notin I_{\sigma\beta}^A[t] = I_\beta^A[\sigma t].$$

(DEF-X):
Here $\bot \notin \{\, \beta(x)\, \} = I_\beta^A[x]$, because of $\beta(x) \neq \bot$.

(DEF-D):
Because of the premises, $\bot \notin I_\beta^A[t1]$, $\;I_\beta^A[t1] \supseteq I_\beta^A[t2]$, so $\bot \notin I_\beta^A[t2]$.

(STR):
Due to the premise $\bot \notin I_\beta^A[f(t_1,\ldots,t_n)]$. Using definition 6.2 (strictness),

$$\bot \notin I_\beta^A[f(t_1,\ldots,t_n)] \text{ implies } \bot \notin I_\beta^A[t_i] \qquad\qquad \lozenge$$

Proof of Theorem 6.9

As in theorem 2.14, valuations in $P\Sigma/R$ are represented by substitutions.

Lemma 6.9.1

$$I_{[\sigma]}^{P\Sigma/R}[t] = \{\, [t'] \mid t' \in W(\Sigma) \wedge \vdash \mathrm{DET}(t') \wedge \vdash \mathrm{DEF}(t') \wedge \vdash \sigma t \to t'\,\}$$

$$\cup \{\, \bot \mid \exists\, t' \in W(\Sigma)\colon \vdash \sigma t \to t' \wedge \vdash \uparrow t'\,\} \quad \text{(for } t \in W(\Sigma, X)).$$

Proof of the lemma:

Induction on the term structure of t, analoguous to lemma 2.14.1:

t=x, x∈X:
$$I^{P\Sigma/R}_{[\sigma]}[t] = \{ [\sigma](x) \} = \{ [\sigma x] \}$$
$$\subseteq \{ [t'] \mid \; \vdash DET(t') \; \wedge \; \vdash DEF(t') \; \wedge \; \vdash \sigma x \rightarrow t' \}.$$

The $\supseteq$-direction holds, too, in analogy to theorem 2.14.

$t = f(t_1,...,t_n) / \subseteq$-direction:

Case 1: $\perp \notin I^{P\Sigma/R}_{[\sigma]}[f(t_1,...,t_n)]$

The property $[t'] \in I^{P\Sigma/R}_{[\sigma]}[f(t_1,...,t_n)]$ implies by definitions 6.2, 6.8

and the induction hypothesis that

$$\exists t_1',...,t_n':$$
$$\vdash DET(t_i') \; \wedge \; \vdash DEF(t_i') \; \wedge \; \vdash \sigma t_i \rightarrow t_i' \; \wedge \; \vdash f(t_1',...,t_n') \rightarrow t'$$
Using (CONG), (TRANS), we have $\vdash \sigma t \rightarrow t'$.

Case 2: $\perp \in I^{P\Sigma/R}_{[\sigma]}[f(t_1,...,t_n)]$

There are again two subcases.

Case 2.1: $\exists \; i \in \{1,...,n\}: \perp \in I^{P\Sigma/R}_{[\sigma]}[t_i]$

For i we can apply the induction hypothesis, giving
$$\exists \; t0: \vdash \sigma t_i \rightarrow t0 \; \wedge \; \vdash \uparrow t0$$
Using (CONG), we have
$$\vdash f(\sigma t_1,...,\sigma t_i,...,\sigma t_n) \rightarrow f(\sigma t_1,...,t0,...,\sigma t_n)$$
If there was a term t1 such that
$$\vdash f(\sigma t_1,...,t0,...,\sigma t_n) \rightarrow t1 \text{ and } \vdash DEF(t1),$$
then, because of the partial DET–completeness and the DET-additivity
there had to be terms $t_1',..., t0',..., t_n'$ such that
$$\vdash f(t_1',...,t0',...,t_n') \rightarrow t1$$
and (besides other facts)
$$\vdash t0 \rightarrow t0' \; \wedge \; \vdash DEF(t0') \; \wedge \; \vdash DET(t0'),$$

which contradicts to the definition of $\vdash \uparrow t0$. So $\vdash \uparrow f(\sigma t_1, \ldots, t0, \ldots, \sigma t_n)$. Using (TRANS), this yields the claim.

Case 2.2:

If the condition for case 2.1 is not fulfilled, the undefined result value must come from the semantics of the function symbol f:
$\exists t_1', \ldots, t_n'$:

$$\vdash DET(t_i') \wedge \vdash DEF(t_i') \wedge \vdash \sigma t_i \to t_i' \wedge \bot \in I_{[\sigma]}^{P\Sigma/R}[f(t_1', \ldots, t_n')]$$

From definition 2.8 follows

$$\exists t0: \vdash f(t_1', \ldots, t_n') \to t0 \wedge \vdash \uparrow t0,$$

which gives using (TRANS), (CONG): $\vdash f(\sigma t_1, \ldots, \sigma t_n) \to t0$.

$\underline{t = f(t_1, \ldots, t_n) / \supseteq\text{-Direction:}}$

Case 1: $\bot \notin I_{[\sigma]}^{P\Sigma/R}[f(t_1, \ldots, t_n)]$

Let $t' \in W(\Sigma)$ such that

$$\vdash DET(t') \wedge \vdash DEF(t') \wedge \vdash f(\sigma t_1, \ldots, \sigma t_n) \to t' .$$

Because of the partial DET-additivity (1) there exist terms $t_1', \ldots, t_n'$ such that

$$\vdash DET(t_i') \wedge \vdash DEF(t_i') \wedge \vdash \sigma t_i \to t_i' \wedge \vdash f(t_1', \ldots, t_n') \to t',$$

that is $t' \in I_{[\sigma]}^{P\Sigma/R}[t]$.

Case 2: $\bot \in I_{[\sigma]}^{P\Sigma/R}[f(t_1, \ldots, t_n)]$

Let $t' \in W(\Sigma)$ such that

$$\vdash f(t_1, \ldots, t_n) \to t' \wedge \vdash \uparrow t'.$$

Because of the partial DET-additivity (2), there exist terms $t_1', \ldots, t_n'$ such that $\vdash f(t_1', \ldots, t_n') \to t'$, that is $\bot \in I_{[\sigma]}^{P\Sigma/R}[t]$.   ◊ (lemma 6.9.1)

For the proof of theorem 6.9, we proceed differently according to the three types of formulas.

Let $\langle DET(t) \rangle \in R$, $[\sigma]$ a valuation. Then, using lemma 6.9.1:

$$I_{[\sigma]}^{P\Sigma/R}[t] = \{ [t'] \mid \vdash DET(t') \wedge \vdash DEF(t') \wedge \vdash \sigma t \to t' \}$$

$$\cup \{ \bot \mid \exists t': \vdash \sigma t \to t' \wedge \vdash \uparrow t' \}$$

If $\perp\notin I_{[\sigma]}^{P\Sigma/R}[t]$, we have as in theorem 2.14: $\mid I_{[\sigma]}^{P\Sigma/R}[t]\mid = 1$.

If $\perp\in I_{[\sigma]}^{P\Sigma/R}[t]$, the rules (AXIOM-2-D) and (DET-R) give

$\vdash$ DET(σt), $\vdash$ t' $\to$ σt.

If there was now a term t'' such that

$\vdash$ DET(t''), $\vdash$ DEF(t'') and $\vdash$ $\sigma t \to$ t'',

this would lead to a contradiction to $\uparrow$t' (because of (TRANS)).

So $I_{[\sigma]}^{P\Sigma/R}[t] = \{\perp\}$, i. e. $\mid I_{[\sigma]}^{P\Sigma/R}[t]\mid = 1$.

Let $\langle$DEF(t)$\rangle \in$R, [σ] a valuation. Then with (AXIOM-3-D) $\vdash$DEF(σt). If now we had $\perp\in I_{[\sigma]}^{P\Sigma/R}[t]$, i. e. according to lemma 4.9.1

$\exists$ t': $\vdash$ $\sigma t \to$ t', $\vdash$ $\uparrow$t',

then using (DEF-D) we could deduce $\vdash$ DEF(t'), in contradiction to the definition of $\uparrow$t'. So $\perp\notin I_{[\sigma]}^{P\Sigma/R}[t]$.

Let $\langle$t1 $\to$ t2$\rangle \in$R, [σ] a valuation. Then analoguously to theorem 2.14:

$$[t'] \in I_{[\sigma]}^{D\Sigma/R}[t2] \Rightarrow [t'] \in I_{[\sigma]}^{D\Sigma/R}[t1] \qquad \text{and}$$

$$\perp\in I_{[\sigma]}^{D\Sigma/R}[t2].+$$

By lemma 6.9.1 this gives

$\exists$ t': $\sigma t2 \to$ t' $\wedge$ $\vdash$ $\uparrow$t'

and by (AXIOM-1-D), (TRANS) $\vdash$ $\sigma t1 \to$ t'. Using lemma 6.9.1 again:

$$\perp\in I_{[\sigma]}^{D\Sigma/R}[t1] \qquad\qquad\qquad \Diamond$$

Proof of Lemma 6.15:

By induction on the term structure of t it can be shown that:

φ is a loose homomorphism

$$\Leftrightarrow \forall\ t\in W(\Sigma): \{e'\in\varphi(e) \mid e\in I^A[t]\backslash\{\perp\}\} \subseteq I^B[t] . \qquad (**)$$

For the proof of the lemma we use the fact (**) from above.

(1):   If φ is total, then $\perp\notin\{e'\in\varphi(e) \mid e\in I^A[t]\backslash\{\perp\}\}$, So "$\Rightarrow$" holds because of (**). The inverse direction holds because of A$\in$PGen(Σ).

(2): If $\perp \in I^A[t]$, the claim can be shown easily by induction on the term structure of t.

So let $\perp \notin I^A[t]$, i. e. $I^A[t] \setminus \{\perp\} = I^A[t]$. From the definition of a tight homomorphism follows that $\perp \notin I^B[t]$, since φ is total. So $I^B[t] \setminus \{\perp\} = I^B[t]$. This means $\perp \in I^B[t] \Rightarrow \perp \in I^A[t]$. The same chain of arguments holds also in inverse direction.

(3): For $e \in I^A[t]$, $e \neq \perp$ because of (**) there is to nothing to show. If $\perp \in I^A[t]$ and φ is weak, from (**) follows that $\{e' \in \varphi(e) \mid e \in I^A[t]\} \subseteq I^B[t]$. Inversely, from this proposition follows the weakness of φ, too.

(4): If $\perp \notin I^A[t]$, there is nothing to show.

So let $\perp \in I^A[t]$. Since φ is strict, the definition of a tight homomorphism yields:

$\{e' \in \varphi(e) \mid e \in I^A[t]\} = \{e' \in \varphi(e) \mid e \in I^A[t] \setminus \{\perp\}\} \cup \{\perp\} = (I^B[t] \setminus \{\perp\}) \cup \{\perp\}$.

This last term is equal to $I^B[t]$ iff $\perp \in I^B[t]$, i. e. if φ is weak.

(5): Follows from (2) and (4).   ◊

Proof of Lemma 6.18:

(1): For $t \in W(\Sigma)$, $A \in PMod(T)$ we have: $A \models \uparrow t \Rightarrow T \vdash \uparrow t$.

(If there was a term t' such that $\vdash t \rightarrow t'$, $\vdash DET(t')$, $\vdash DEF(t')$, then according to theorem 6.6. we had $I^A[t] \neq \{\perp\}$, in contradiction to $A \models \uparrow t$.)

So we still have to show that for a minimally defined A holds: $\vdash \uparrow t \Rightarrow A \models \uparrow t$. According to theorem 6.9 $P\Sigma/R \in PMod(T)$. From $\vdash \uparrow t$ follows $P\Sigma/R \models \uparrow t$, so (since A is minimally defined) $A \models \uparrow t$.

(2): Let $e \in s^A$. Since A is term-generated, there is a term $t' \in W(\Sigma)$ such that $I^A[t'] = \{e\}$. So $\neg(A \models \uparrow t')$. According to part (1) of the lemma follows $\neg(T \vdash \uparrow t)'$. This means by definition 6.7: $\exists$ t: $\vdash t' \rightarrow t \wedge \vdash DEF(t)$. Using theorem 6.6 follows $I^A[t] = \{e\}$.

(3): Let $e \in s^A$. According to part (2) of the lemma there is a $t \in W(\Sigma)$ with $T \vdash DEF(t) \wedge I^A[t] = \{e\}$. Because of the partial DET-completenenss of T there is now a t' with $\vdash t \rightarrow t' \wedge \vdash DET(t') \wedge \vdash DEF(t')$. According to theorem 6.6 follows $I^A[t'] = \{e\}$. The rest of the proof is in complete analogy to lemma 3.12.   ◊

Proof of Lemma 6.21:

(1) ⇒ (2):
Let A be partially maximally deterministic, B∈PGen(T), φ: B → A loose and total. Then for t∈W(Σ) we have the following facts (since A is partially more deterministic than B):
$$|\{e'∈φ(e) \mid e∈I^B[t]\backslash\{\bot\}\}| \geq |\ I^B[t]\backslash\{\bot\}\ | \geq |\ I^A[t]\backslash\{\bot\}\ |.$$
Using lemma 6.15 (1):
$$\{e'∈φ(e) \mid e∈I^B[t]\backslash\{\bot\}\} \subseteq I^A[t]\backslash\{\bot\}, \text{ and so}$$
$$\{e'∈φ(e) \mid e∈I^B[t]\backslash\{\bot\}\} = I^A[t]\backslash\{\bot\}.$$
Since A is totally more deterministic than B, we have $\bot∈I^A[t] \Rightarrow \bot∈I^B[t]$. So from lemma 6.15 (2) it follows that φ is tight.

(2) ⇒ (3):
The continuation φ of I^A is a loose total homomorphism from PΣ/R to A. According to the precondition, φ is tight, i. e. with lemma 6.15 (2) :
$$\{\ e∈I^A[t'] \mid \vdash t{\to}t' \wedge \vdash DEF(t') \wedge \vdash DET(t')\ \} = I^A[t]\backslash\{\bot\}\ .$$
So for any $e∈I^A[t]$, $e{\neq}\bot$, there is a term t' such that
$$\vdash t{\to}t' \wedge \vdash DEF(t') \wedge \vdash DET(t')\ .$$
Here $|\ I^A[t']\ | = 1$ and (because of theorem 6.6) $e∈I^A[t']$, i. e. $I^A[t'] = \{\ e\ \}$.
If $\bot∈I^A[t]$, according to lemma 6.15 (2): $\bot∈I^{PΣ/R}[t]$, i. e.
$$\exists\ t': \vdash t \to t'\ \wedge\ \vdash \uparrow t'.$$
Since φ is a tight homomorphism, we have:
$$\{\ e∈I^A[t''] \mid \vdash t'{\to}t' \wedge \vdash DEF(t'') \wedge \vdash DET(t'')\ \} = I^A[t']\backslash\{\bot\} = \emptyset$$
(the last equation because of $\vdash \uparrow t'$). So $I^A[t'] = \{\bot\}$.

(3) ⇒ (1):
Let B be a refinement of A, i. e. B∈PGen(T), φ: B ⇀ A a total loose homomorphism, t∈W(Σ).
For $e∈I^A[t]\backslash\{\bot\}$, according to the precondition, there is a term t' such that
$$I^A[t'] = \{\ e\ \} \wedge \vdash t{\to}t' \wedge \vdash DEF(t') \wedge \vdash DET(t').$$
So there is an $e'∈s^B$ with
$$I^B[t'] = \{\ e'\ \}\wedge\ e'{\neq}\bot \wedge e'∈I^B[t]\ .$$

The homomorphism condition gives $\varphi(e') = \{e\}$. So φ is a surjective pointwise mapping from $I^B[t]\backslash\{\perp\}$ to $I^A[t]\backslash\{\perp\}$, i. e. $|I^B[t]\backslash\{\perp\}| \geq |I^A[t]\backslash\{\perp\}|$.

Moreover, due to the precondition
$$\perp \in I^A[t] \Rightarrow \quad \exists\, t': |\text{-}\, t \rightarrow t' \,\wedge\, |\text{-}\uparrow t' \,\wedge\, I^A[t'] = \{\perp\}.$$
The homomorphism condition for t' gives:
$$\{e' \in \varphi(e) \mid e \in I^B[t']\backslash\{\perp\}\} \subseteq I^A[t']\backslash\{\perp\} = \emptyset.$$
This is only possible (since $I^B[t'] \neq \emptyset$), if $I^B[t'] = \{\perp\}$. So $\perp \in I^B[t]$.   ◊

Proof of Theorem 6.27:

Let $A \in DPGen(T)$. The proof uses the following lemma:

Lemma 6.27.1

$$E(R)\, |\text{-}\, t1 \rightarrow t2 \,\wedge\, R\, |\text{-}\, t2 \rightarrow t3 \,\wedge\, |\text{-}\, DET(t3) \,\wedge\, |\text{-}\, DEF(t3) \;\Rightarrow$$
$$\exists\, e \in I^A[t1], t' \in W(\Sigma):$$
$$I^A[t'] = \{e\} \,\wedge\, t' \leftrightarrow t3 \,\wedge\, |\text{-}\, DET(t') \,\wedge\, |\text{-}\, DEF(t')$$

Proof of the lemma:

By induction on the deduction of $E(R)\, |\text{-}\, t1 \rightarrow t2$.

(REFL):
In this case we have $t1 = t2$. Choose $t' = t3$ and accordingly $I^A[t3] = \{e\}$.

(TRANS):
There is a t0 with $E(R)\, |\text{-}\, t1 \rightarrow t0$ and $E(R)\, |\text{-}\, t0 \rightarrow t2$.
According to the induction hypothesis there are $e0 \in I^A[t0]$, t0' with $t0' \leftrightarrow t3$, $|\text{-}\, DEF(t0')$, $|\text{-}\, DET(t0')$. By lemma 6.21 (3) there is a t0'' such that
$$R\, |\text{-}\, t0 \rightarrow t0'' \,\wedge\, |\text{-}\, DEF(t0'') \,\wedge\, |\text{-}\, DET(t0'') \,\wedge\, I^A[t0''] = \{e0\}.$$
Using definition 6.25 we have now $t0' \leftrightarrow t0''$.
According to the induction hypothesis there are $e \in I^A[t1]$, t' with
$$I^A[t'] = \{e\}, t' \leftrightarrow t0'', |\text{-}\, DET(t'), |\text{-}\, DEF(t').$$
The transitivity of $\leftrightarrow$ gives $t' \leftrightarrow t3$.

(CONG):

Let $t1 = f(t_1,...,t_i,...,t_n)$, $t2 = f(t_1,...,t_i',...,t_n)$, $E(R)$ ⊢ $t_i \rightarrow t_i'$. Since T is partially DET-additive, there are $t_1'',...,t_i'',...,t_n''$ with

$\quad$ ⊢ $DET(t_j'')$, ⊢ $DEF(t_j'')$,

$\quad$ R ⊢ $t_1 \rightarrow t_1''$, ..., R ⊢ $t_i' \rightarrow t_i''$, ..., R ⊢ $t_n \rightarrow t_n'$, R ⊢ $f(t_1'',...,t_n'') \rightarrow t3$.

So the induction hypothesis can be applied. Therefore there are $e' \in I^A[t_i]$, t''

with $\quad I^A[t''] = \{e'\}$, $t'' \Leftrightarrow t_i''$, ⊢ $DEF(t'')$, ⊢ $DET(t'')$.

Because of R ⊢ $f(t_1'',...,t_n'') \rightarrow t3$ and $t'' \Leftrightarrow t_i''$ there are B_1, ..., $B_{k+1} \in DPGen(T)$ where

$\quad B_1|= f(t_1'',...,t'',...,t_n'') \rightarrow u_1$, ..., $B_{k+1}|= u_k \rightarrow t3$.

By lemma 6.21 (3) we have

$\quad$ R ⊢ $u_k \rightarrow t3$, ..., R ⊢ $f(t_1'',...,t'',...,t_n'') \rightarrow t3$.

Let now $t' = t3$. Then, trivially,

$\quad t' \Leftrightarrow t3$, ⊢ $DET(t')$, ⊢ $DEF(t')$.

Moreover

$\quad I^A[t3] \subseteq I^A f(t_1'',...,t'',...,t_n'')$

(because of R ⊢ $f(t_1'',...,t'',...,t_n'') \rightarrow t3$) and for $i \neq j$ $I^A[t_j''] \subseteq I^A[t_j]$ (because of R ⊢ $t_j \rightarrow t_j'$) and $I^A[t''] \subseteq I^A[t_i]$ (consequence of the induction hypothesis). Using the additivity of f^A:

$\quad I^A[t3] \subseteq I^A[f(t_1,...,t_n)]$.

(AXIOM-1-D):

We distinguish according to the axiom applied.

Case 1: Application of an axiom $\triangleleft l \rightarrow r \triangleright \in R$

Then R ⊢ $t1 \rightarrow t2$, and so we can choose $t' = t3$.

Case 2: Application of an axiom $\triangleleft l \rightarrow r \triangleright \in E(R) \backslash R$

Then $t1 \Leftrightarrow t2$, i. e. ⊢ $DET(t1)$, ⊢ $DET(t2)$, and therefore $t2 \Leftrightarrow t3$. We can choose $t' = t2$, and e such that $I^A[t2] = \{e\}$. $\qquad \lozenge$ (lemma 6.27.1)

For the proof of theorem 6.27, we define the needed homomorphism by

$\quad \varphi(e) = \{[t]\} \qquad$ such that $I^A[t] = \{e\}$ ∧ ⊢ $DEF(t)$ ∧ ⊢ $DET(e)$.

The existence of such a t follows from the partial DET-completeness and from the fact that A is term-generated. The uniqueness follows from the definition of $E(R)$:

$\quad I^A[t'] = \{e\}$ ∧ ⊢ $DEF(t')$ ∧ ⊢ $DET(t')$

$\Rightarrow$ $E(R) \vdash t \rightarrow t'$

$\Rightarrow$ $[t] = [t']$.

By theorem 6.22 we have:

$\bot \in I^A[t]$

$\Leftrightarrow$ $\exists t': R \vdash t \rightarrow t' \land R \vdash \uparrow t'$

$\Leftrightarrow$ $\exists t': E(R) \vdash t \rightarrow t' \land E(R) \vdash \uparrow t'$ (since $R \vdash DEF(t) \Leftrightarrow E(R) \vdash DEF(t)$)

$\Leftrightarrow$ $\bot \in I^{P\Sigma/E(R)}[t]$.

For the condition of a strong homomorphism in lemma 6.15 (5) we have to show:

$$\{[t'] \in \varphi(e) \mid e \in I^A[t] \backslash \{\bot\}\} = I^{P\Sigma/E(R)} \backslash \{\bot\}.$$

Let $[t'] \in \varphi(e)$, i. e. $I^A[t'] = \{e\}$ and $e \in I^A[t]$. According to lemma 6.21 (3) there is a t'' such that $I^A[t''] = \{e\}$ and $R \vdash t \rightarrow t''$, so because of the definition of $E(R)$: $E(R) \vdash t'' \rightarrow t$ and $E(R) \vdash t \rightarrow t''$. So $[t''] \in I^{P\Sigma/E(R)}[t]$.

Let $[t'] \in I^{P\Sigma/E(R)}[t]$, i. e. $E(R) \vdash t \rightarrow t'$, $\vdash DEF(t')$, $\vdash DET(t')$. By lemma 6.27.1 there are $e \in I^A[t]$, t'' such that $I^A[t''] = \{e\}$ and $[t''] = [t']$, i. e. $[t'] \in \varphi(e)$. $\Diamond$

Appendix B: Experiments with RAP

This appendix shows how the example specification from chapter 7 can be used to obtain directly a running (but not very efficient) interpreter for CP.

B.1. General Remarks

The experiments reported here have been performed with RAP Version 3.0 on a SUN SPARCstation 10 computer with 32 MByte RAM.

The syntax of RAP specifications (*types*) is very similar to the syntax used in this book. In order to improve the readability of terms, RAP is able to display terms in a so-called *mixfix*-notation. For instance for the operation prefix the notation:

```
    _ -> _:  (Com,Agent)Agent
```
can be used instead of

        prefix: Com × Agent → Agent .

The term

        prefix(a,prefix(b,stop))

in this case is shown as like

```
        a -> b -> stop.
```
Also the empty string can be used to represent an operation (see for instance the operation generating an empty set in COM_SET).

Texts printed in a `typewriter font` are original in- or output of the RAP system (except of some brackets which have been inserted at a few places by hand to improve the readability).

In general, all specifications have been formulated in a constructor-oriented style, since RAP supports the notion of a constructor. As it was shown above, RAP then automatically respects the DET-axioms for the constructor operations (and DET-axioms then are superfluous).

B.2. Specifications (types)

The specification COM is restricted to three elementary communication actions, in order to keep the search space small:

```
type COM
  basedon BOOL

  sort Com
  cons a: Com,    b: Com,    c: Com
  func _ == _: (Com,Com)Bool

axioms all (i: Com)

  (1) i == i -> true,        (2) a == b -> false,
  (3) a == c -> false,       (4) b == a -> false,
  (5) b == c -> false,       (6) c == a -> false,
  (7) c == b -> false

endoftype
```

COM_SET specifies finite sets of communication actions. Sets are here implemented by sequences, since axioms for commutativity, associativity or idempotence (for insert) would violate the restrictions of constructor-based specifications. The operation _[_]_ describes the "majority"-operator for sets which is needed later on.

```
type COM_SET
  basedon COM,BOOL

  sort ComSet
  cons  : ComSet, {empty set as empty symbol}
        _ _: (ComSet,Com)ComSet

  func _ IN _: (Com,ComSet)Bool,
        _[_]_: (ComSet,ComSet,ComSet)ComSet

axioms all (i, j: Com, s, s1, s2: ComSet)

  (1) i IN  -> false,
  (2) i IN s j -> i == j or (i IN s),
  (3) []s -> ,
  (4) j IN s2 -> true =>  s1 j[] s2 -> s1[]s2 j,
  (5) j IN s2 -> false => s1 j[] s2 -> s1[]s2,
  (6) j IN s1 -> true =>  s1[s j]s2 -> s1[s]s2 j,
  (7) j IN s2 -> true =>  s1[s j]s2 -> s1[s]s2 j,
```

```
    (8) j IN s1 -> false & j IN s2 = false =>
                                    s1[s j]s2 -> s1[s]s2
  endoftype
```

The specification AGENT is identical to the version in chapter 7:

```
type AGENT
  basedon ID,COM,COM_SET

  sort Agent
  cons STOP: Agent,    DIV: Agent,
       _ -> _: (Com,Agent)Agent,
       (_ OR _): (Agent,Agent)Agent,
       (_ [] _): (Agent,Agent)Agent,
       (_ ||{_} _): (Agent,ComSet,Agent)Agent,
       (_ :: _): (Id,Agent)Agent,
       _: (Id)Agent

endoftype
```

The specification PAIR basically introduces a Cartesian product between Action and Agent. Moreover it admits a special element (LOCK), which is used below in STEP for the totalization with respect to deadlock:

```
type PAIR
  basedon COM,AGENT

  sort Pair
  cons <_,_>: (Com,Agent)Pair,    LOCK: Pair

endoftype
```

The auxiliary specification SUBST provides a syntactical substitution operation on the sort Agent:

```
type SUBST
  basedon ID,COM,BOOL,AGENT,COM_SET

  func _[_/_]: (Agent,Id,Agent)Agent

axioms all (i, j: Id, p, q, q1, q2: Agent,
            A: ComSet, x: Com)

    (1)  STOP[i/p] -> STOP,
    (2)  DIV[i/p] -> DIV,
    (3)  (x -> q[i/p]) -> (x -> q[i/p]),
    (4)  (q1 OR q2)[i/p] -> (q1[i/p] OR q2[i/p]),
```

```
(5)   (q1 [] q2)[i/p] -> (q1[i/p] [] q2[i/p]),
(6)   (q1 ||{A} q2)[i/p] ->
         (q1[i/p] ||{A} q2[i/p]),
(7)   (i == j) -> true => j[i/p] -> p,
(8)   (i == j) -> false => j[i/p] -> j,
(9)   (i == j) -> true =>
         (j :: q)[i/p] -> (j :: q),
(10)  (i == j) -> false =>
         (j :: q)[i/p] -> (j :: q[i/p])

endoftype
```

STEP contains, compared to chapter 7, additional axioms for the totalization in
the case of deadlock:

```
type STEP
   basedon AGENT,COM,COM_SET,PAIR,SUBST,BOOL,ID

   func step(_): (Agent)Pair

axioms all (i, j: Com, p, p1, q, q1: Agent,
            A: ComSet, x: Id)

   (STOP)    step(STOP) -> LOCK,
   (PREFIX)  step(i -> p) -> <i,p>,
   (OR1)     step((p OR q)) -> step(p),
   (OR2)     step((p OR q)) -> step(q),
   (CHC1)    step((p [] q)) -> step(p),
   (CHC2)    step((p [] q)) -> step(q),
   (PAR1)    step(p) -> <i,p1> & step(q) -> <j,q1> &
             (i IN A) -> true & i == j -> true =>
                step((p ||{A} q)) -> <i,(p1 ||{A} q1)>,
   (PAR2)    step(p) -> <i,p1> & step(q) -> <j,q1> &
             (i IN A) -> true & (j IN A) -> true &
             i == j -> false =>
                step((p ||{A} q)) -> LOCK,
   (PAR3)    step(p) -> <i,p1> & (i IN A) -> false =>
                step((p ||{A} q)) -> <i,(p1 ||{A} q)>,
   (PAR4)    step(q) -> <i,q1> & (i IN A) -> false =>
                step((p ||{A} q)) -> <i,(p ||{A} q1)>,
   (REC)     step((x :: p)) -> step(p[x/(x :: p)])

endoftype
```

The trace and refusal semantics now can be defined like in chapter 7.

```
type TRACE
  basedon COM,AGENT,PAIR,STEP

  sort Trace
  cons <>: Trace,
       _._: (Com,Trace)Trace

  func trace(_): (Agent)Trace

axioms all (p, p1: Agent, i: Com)

  (TRC1) trace(p) -> <>,
  (TRC2) step(p) -> <i,p1> =>
         trace(p) -> i.trace(p1)

endoftype


type REFUSE
  basedon COM,COM_SET,AGENT,SUBST,ID,BOOL

  func refuse(_): (Agent)ComSet

axioms all (p, q: Agent, i: Com, A, M: ComSet,
           x: Id)

  (REF_STOP)   refuse(STOP) -> M,
  (REF_PREFIX) i IN M -> false =>
                 refuse(i -> p) -> M,
  (REF_OR1)    refuse((p OR q)) -> refuse(p),
  (REF_OR2)    refuse((p OR q)) -> refuse(q),
  (REF_CHC)    refuse(p) -> M & refuse(q) -> M =>
                 refuse((p [] q)) -> M,
  (REF_PAR)    refuse((p ||{A} q)) ->
                 refuse(p)[A]refuse(q),
  (REF_REC)    refuse((x :: p)) ->
                 refuse(p[x/(x :: p)])

endoftype
```

B.3. Experiments (tasks)

The RAP system is able to enumerate solutions for a system of equations based on a given specification. This mechanism can be used to simulate processes in

CP. Below for each experiment the equation to be solved is given (x is the unknown variable); then the solutions found by RAP are reported, giving also the approximative CPU time needed.

The first agent consists of two (sequential) parts which run in parallel, without any synchronization. The traces consist here in all so-called *interleavings* of both processes.

```
trace((a->b->STOP ||{} b->a->STOP)) = x
```

13 solutions found.

```
[x = <>]
[x = a.<>]
[x = b.<>]
[x = a.b.<>]
[x = b.a.<>]
[x = a.b.b.<>]
[x = a.b.a.<>]
[x = b.a.b.<>]
[x = b.a.a.<>]
[x = a.b.b.a.<>]
[x = a.b.a.b.<>]
[x = b.a.b.a.<>]
[x = b.a.a.b.<>]
```

CPU time: 0.38 secs

The subprocesses from above can be forced to synchronize in one particular action (for instance the action a). Then the whole system behaves sequentially:

```
trace((a->b->STOP ||{ a} b->a->STOP)) = x
```

4 solutions found.

```
[x = <>]
[x = b.<>]
[x = b.a.<>]
[x = b.a.b.<>]
```

CPU time: 0.12 secs

If synchronization in both actions is required, only the trivial trace exists (deadlock):

```
    trace((a ->b->STOP ||{ a b} b->a->STOP)) = x
```

1 solution found.

[x = <>]

CPU time: 0.05 secs

The following examples address the distiction between internal and external nondeterminism. First, an example for the traces of a system of processes which uses **OR**:

```
    trace(((a->STOP OR b->STOP) ||{ a b} a->STOP)) = x
```

2 solutions found.

[x = <>]
[x = a.<>]

CPU time: 0.08 secs

The same system using []: As expected, the same set of traces is computed:

```
    trace(((a->STOP [] b->STOP) ||{ a b} a->STOP)) = x
```

2 solutions found.

[x = <>]
[x = a.<>]

CPU time: 0.10 secs

The refusal remantics can distinguish between the processes. Below the refusal sets for both variants of the example (using OR and choice, respectively) are given, as they are computed by RAP. Both enumerations do not terminate (and have been terminated by user interaction). Please remember that the set of communication actions has been fixed to { a, b, c }:

```
    refuse((a->STOP OR b-> TOP)) = x
```

33 solutions found.

```
[x = ]            [x =  b]          [x =   c]
[x =   a]         [x =  b b]        [x =   c b]
[x =   b c]       [x =  c c]        [x =   a a]
[x =   c a]       [x =  a c]        [x =   b b b]
```

```
[x =  c b b]        [x =  b c b]        [x =  c c b]
[x =  b b c]        [x =  c b c]        [x =  b c c]
[x =  c c c]        [x =  a a a]        [x =  c a a]
[x =  a c a]        [x =  c c a]        [x =  a a c]
[x =  c a c]        [x =  a c c]        [x =  b b b b]
[x =  c b b b]      [x =  b c b b]      [x =  c c b b]
[x =  b b c b]      [x =  c b c b]      [x =  b c c b]

CPU time: 0.50 secs


    refuse((a->STOP [] b->STOP)) = x

5 solutions found.

[x = ]
[x =  c]
[x =  c c]
[x =  c c c]
[x =  c c c c]

CPU time: 0.83 secs
```

This example clearly shows a disadvantage of the set implementation (by sequences) used here: The same set is printed many times using various equivalent representations.

The following examples demonstrate that the trace semantics (with the artificial totalization with respect to deadlock) can distinguish between **stop** and **div**: The enumeration of the traces of **div** does not terminate. But also the refusal semantics is able to distinguish between both processes:

```
    trace(STOP) = x

1 solution found.

[x = <>]

CPU time: 0.02 secs


    trace(DIV) = x

1 solution found.
```

```
[x = <>]
```

Aborted by time limit. (i.e. nontermination)

```
  refuse(STOP) = x
```

1 solution found.

```
[x = *0]
```

CPU time: 0.02 secs

```
  refuse(DIV) = x
```

No solutions found.

CPU time: 0.02 secs

The term *0 in the output above is a system-generated variable which stands for an arbitrary set of communication actions.

Concludingly, the results of a larger example shall be presented, in order to give an impression of the current state of (in-)efficiency of the RAP-generated interpreter for CP. We give a CP process for the famous problem of the "Dining Philosophers" by E. W. Dijkstra. For the sake of simplicity, only *two* philosophers are considered here, which sit at the opposite sides of a table. There are only two forks for the philosophers to eat their meals. The philosophers are called below P1 and P2, the forks F1 and F2. Possible actions are here (leading to a suitable type COM):

> pic(Pi,Fj)   Philosopher i picks up fork j
> put(Pi,Fj)   Philosopher i puts back fork j.

The following equation contains recursive processes for the philosophers and for the forks. Each of the forks is represented by a recursive process (with labels 'f1 and 'f2); each of the philosophers is represented by another recursive process (with labels 'p1 and 'p2). The two fork processes run in parallel without any synchronization, giving a compund fork process. The two philosopher processes run in parrallel without any synchronization, giving a compund philosopher process. The compund fork and the compund philosopher process run in parallel, synchronized by all possible pic- and put-actions.

```
trace(
(
    (      ('f1::(pic(P1,F1)->put(P1,F1) ->'f1
                [] pic(P2,F1)->put(P2,F1) ->'f1))
    ||{}
           ('f2::(pic(P1,F2)->put(P1,F2) ->'f2
                [] pic(P2,F2)->put(P2,F2) ->'f2))
    )
||{pic(P1,F1) pic(P1,F2) pic(P2,F1) pic(P2,F2)
   put(P1,F1) put(P1,F2) put(P2,F1) put(P2,F2)}
    (      ('p1::pic(P1,F1)->pic(P1,F2)->
                put(P1,F1)->put(P1,F2)->'p1)
    ||{} ('p2::pic(P2,F2)->pic(P2,F1)->
                put(P2,F2)->put(P2,F1)->'p2)
    )
)) = x
```

The set of traces computed by RAP is infinite; the enumeration has been stopped again by user interaction. Please note that there are some finite traces (for instance

```
        pic(P1,F1).pic(P2,F2).<>),
```

which are not "continued", i.e., which do no not appear as a non-trivial prefix of any other trace. These traces correspond to the situations where the philosophers die of starvation.

```
23 solutions found.

[x = <>]
[x = pic(P1,F1).<>]
[x = pic(P2,F2).<>]
[x = pic(P1,F1).pic(P1,F2).<>]
[x = pic(P2,F2).pic(P2,F1).<>]
[x = pic(P1,F1).pic(P2,F2).<>]
[x = pic(P2,F2).pic(P1,F1).<>]
[x = pic(P1,F1).pic(P1,F2).put(P1,F1).<>]
[x = pic(P2,F2).pic(P2,F1).put(P2,F2).<>]
[x = pic(P1,F1).pic(P1,F2).put(P1,F1).put(P1,F2).<>]
[x = pic(P2,F2).pic(P2,F1).put(P2,F2).put(P2,F1).<>]
[x = pic(P1,F1).pic(P1,F2).put(P1,F1).put(P1,F2).
        pic(P1,F1).<>]
[x = pic(P1,F1).pic(P1,F2).put(P1,F1).put(P1,F2).
        pic(P2,F2).<>]
[x = pic(P2,F2).pic(P2,F1).put(P2,F2).put(P2,F1).
        pic(P1,F1).<>]
[x = pic(P2,F2).pic(P2,F1).put(P2,F2).put(P2,F1).
        pic(P2,F2).<>]
```

```
[x = pic(P1,F1).pic(P1,F2).put(P1,F1).put(P1,F2).
        pic(P1,F1).
     pic(P1,F2).<>]
[x = pic(P1,F1).pic(P1,F2).put(P1,F1).put(P1,F2).
        pic(P2,F2).pic(P2,F1).<>]
[x = pic(P2,F2).pic(P2,F1).put(P2,F2).put(P2,F1).
        pic(P1,F1).pic(P1,F2).<>]
[x = pic(P2,F2).pic(P2,F1).put(P2,F2).put(P2,F1).
        pic(P2,F2).pic(P2,F1).<>]
[x = pic(P1,F1).pic(P1,F2).put(P1,F1).put(P1,F2).
        pic(P1,F1).pic(P2,F2).<>]
[x = pic(P1,F1).pic(P1,F2).put(P1,F1).put(P1,F2).
        pic(P2,F2).pic(P1,F1).<>]
[x = pic(P2,F2).pic(P2,F1).put(P2,F2).put(P2,F1).
        pic(P1,F1).pic(P2,F2).<>]
[x = pic(P2,F2).pic(P2,F1).put(P2,F2).put(P2,F1).
        pic(P2,F2).pic(P1,F1).<>]
```

CPU time: 20.33 secs

It is worth noting that the same experiment in 1988 (with RAP 2.1 on a SUN 3/160 computer) needed more than 500 seconds of CPU time. It is obvious that this gain in performance will help in the process of bringing "very high level programming languages" like RAP into practice.

Progress in Theoretical Computer Science

Progress in Theoretical Computer Science is a series that focuses on the theoretical aspects of computer science and on the logical and mathematical foundations of computer science, as well as the applications of computer theory. It addresses itself to research workers and graduate students in computer and information science departments and research laboratories, as well as to departments of mathematics and electrical engineering where an interest in computer theory is found.

The series publishes research monographs, graduate texts, and polished lectures from seminars and lecture series. We encourage preparation of manuscripts in some form of TeX for delivery in camera-ready copy, which leads to rapid publication, or in electronic form for interfacing with laser printers or typesetters.

Proposals should be sent directly to the Editor, any member of the Editorial Board, or to: Birkhäuser Boston, 675 Massachusetts Ave., Cambridge, MA 02139. The Series includes: